T0226091

Kernel Methods for Machine Learning with Math and R

Joe Suzuki

Kernel Methods for Machine Learning with Math and R

100 Exercises for Building Logic

 Springer

Joe Suzuki
Graduate School of Engineering Science
Osaka University
Toyonaka, Osaka, Japan

ISBN 978-981-19-0397-7   ISBN 978-981-19-0398-4   (eBook)
https://doi.org/10.1007/978-981-19-0398-4

This Springer imprint is published by the registered company Springer Nature Singapore Pte Ltd.
The registered company address is: 152 Beach Road, #21-01/04 Gateway East, Singapore 189721,
Singapore

Preface

How to Overcome Your Kernel Weakness

Among machine learning methods, kernels have always been a particular weakness of mine. I tried to read "Introduction to Kernel Methods" by Kenji Fukumizu (in Japanese) but failed many times. I invited Prof. Fukumizu to give an intensive lecture at Osaka University and listened to the course for a week with the students, but I could not understand the book's essence. When I first started writing this book, my goal was to rid my sense of weakness. However, now that this book is completed, I can tell readers how they can overcome their own kernel weaknesses.

Most people, even machine learning researchers, do not understand kernels and use them. If you open this page, I believe you have a positive feeling that you want to overcome your weakness.

The shortest path I would most recommend for achieving this is to learn mathematics by starting from the basics. Kernels work according to the mathematics behind them. It is essential to think through this concept until you understand it. The mathematics needed to understand kernels are called functional analysis (Chap. 2). Even if you know linear algebra or differential and integral calculus, you may be confused. Vectors are finite dimensional, but a set of functions is infinite dimensional and can be treated as linear algebra. If the concept of completeness is new to you, I hope you will take the time to learn about it. However, if you get through this second chapter, I think you will understand everything about kernels.

This book is the third volume (of six) in the 100 Exercises for Building Logic set. Since this is a book, there must be a reason for publishing it (the so-called cause) when existing books on kernels can be found. The following are some of the features of this book.

1. The mathematical propositions of kernels are proven, and the correct conclusions are stated so that the reader can reach the essence of kernels.

2. As in the other books in the 100 Mathematical Problems in Machine Learning series, source programs and running examples are presented to promote understanding. It is not easy for readers to understand the results if only mathematical formulas are given, and this is especially true for kernels.
3. Once the reader understands the basic topics of functional analysis (Chap. 2), the applications in the subsequent chapters are discussed, and no prior knowledge of mathematics is assumed.
4. This kernel considers both the kernel of the RKHS and the kernel of the Gaussian process. A clear distinction is made between the two treatments. In this book, the two types of kernels are discussed in Chaps. 5 and 6, respectively.

We surveyed books on kernels both in Japan and overseas but found that none satisfied two or more of the above characteristics.

I have experienced many failures leading up to the publication of this book. Every year, I give a lecture (at the graduate school of Osaka University). Each area of machine learning is studied by solving 100 mathematical and programming exercises. Sparse estimation (2018) and graphical models (2019) have gained popularity, and the 2020 kernel lecture has more than 100 students enrolled. However, although I prepared for the lectures for more than 2 days every week, the talks did not go well, probably due to my weakness regarding the subject. This was evident from the class questionnaires provided by the students. However, I analyzed each of these problems and made improvements, and this book was born.

I hope that readers will learn about kernels efficiently without following the same path that I took (consuming much time and energy through trial and error). Reading this book does not mean that you will write a paper immediately, but it will give you a solid foundation. You will be able to read kernel papers smoothly, which had previously seemed difficult, and you will be able to see the whole kernel paradigm from a higher level. This book is also enjoyable, even for researchers in machine learning. We hope that you will use this book to achieve success in your respective fields.

What Makes KMMR Unique?

1. Developing logic
    We mathematically formulate and solve each ML problem and build those programs to grasp the subject's essence. The KMMR (Kernel methods for Machine learning with Math and R) instills "logic" in the minds of the readers. The reader will acquire both the knowledge and ideas of ML. Even if new technology emerges, they will be able to follow the changes smoothly. After solving the 100 problems, most students would say, "I learned a lot".
2. Not just a story
    If programming codes are available, you can immediately take action. It is unfortunate when an ML book does not offer the source codes. Even if a package

is available, if we cannot see the inner workings of the programs, all we can do is input data into those programs. In KMMR, the program codes are available for most of the procedures. In cases where the reader does not understand the math, the codes will help them know what it means.

3. Not just a how-to book: an academic book written by a university professor
This book explains how to use the package and provides examples of executions for those unfamiliar with them. Still, because only the inputs and outputs are visible, we can see the procedure as a black box. In this sense, the reader will have limited satisfaction because they will not obtain the subject's essence. KMMR intends to show the reader the heart of ML and is more of a full-fledged academic book.

4. Solve 100 exercises: problems are improved with feedback from university students
The exercises in this book have been used in university lectures and refined based on students' feedback. The best 100 problems were selected. Each chapter (except the exercises) explains the solutions, and you can solve all of the exercises by reading the book.

5. Self-contained
All of us have been discouraged by phrases such as "for the details, please refer to the literature XX". Unless you are an enthusiastic reader or researcher, nobody will seek out those references. In this book, we have presented the material so that consulting external references is not required. Additionally, the proofs are simple derivations, and the complicated proofs are given in the appendices at the end of each chapter. KMMR completes all discussions, including the appendices.

6. Readers' pages: questions, discussion, and program files
The reader can ask any question on the book via https://bayesnet.org/books.

Osaka, Japan
November 2021

Joe Suzuki

Acknowledgments

The author wishes to thank Mr. Bing Yuan Zhang, Mr. Tian Le Yang, Mr. Ryosuke Shimmura, Mr. Tomohiro Kamei, Ms. Rieko Tasaka, Mr. Keito Odajima, Mr. Daiki Fujii, Mr. Hongming Huang, and all graduate students at Osaka University, for pointing out logical errors in mathematical expressions and programs. Furthermore, I would like to take this opportunity to thank Dr. Hidetoshi Matsui (Shiga University), Dr. Michio Yamamoto (Okayama University), and Dr. Yoshikazu Terada (Osaka University) for their advice on functional data analysis in seminars and workshops. This English book is based mainly on the Japanese book published by Kyoritsu Shuppan Co., Ltd. in 2021. The author would like to thank Kyoritsu Shuppan Co., Ltd., particularly its editorial members Mr. Tetsuya Ishii and Ms. Saki Otani. The author also appreciates Ms. Mio Sugino, Springer, preparing the publication and providing advice on the manuscript.

Osaka Japan
November 2021

Joe Suzuki

Contents

Chapter 1
Positive Definite Kernels

In data analysis and various information processing tasks, we use kernels to evaluate the similarities between pairs of objects. In this book, we deal with mathematically defined kernels called positive definite kernels. Let the elements x, y of a set E correspond to the elements (functions) $\Psi(x)$, $\Psi(y)$ of a linear space H called the reproducing kernel Hilbert space. The kernel $k(x, y)$ corresponds to the inner product $\langle \Psi(x), \Psi(y) \rangle_H$ in the linear space H. Additionally, by choosing a nonlinear map Ψ, this kernel can be applied to various problems. The set E may be a string, a tree, or a graph, even if it is not a real-numbered vector, as long as the kernel satisfies positive definiteness. After defining probability and Lebesgue integrals in the second half, we will learn about kernels by using characteristic functions (Bochner's theorem).

1.1 Positive Definiteness of a Matrix

Let $n \geq 1$; we say that a square matrix A is symmetric if $A \in \mathbb{R}^{n \times n}$ is equal to its transpose $(A^\top = A)$,[1] and we say that A is nonnegative definite if all the eigenvalues are nonnegative.

Proposition 1 (nonnegative definite matrix) *The following three conditions are equivalent for a symmetric matrix $A \in \mathbb{R}^{n \times n}$:*

1. *A matrix $B \in \mathbb{R}^{n \times n}$ exists such that $A = B^\top B$.*
2. *$x^\top A x \geq 0$ for any $x \in \mathbb{R}^n$.*
3. *The eigenvalues of A are nonnegative.*

Proof. $1. \Rightarrow 2.$ holds because $A = B^\top B \Rightarrow x^\top A x = x^\top B^\top B x = \|Bx\|^2 \geq 0. \, 2. \Rightarrow 3.$ follows from the fact that $x^\top A x \geq 0$, $x \in \mathbb{R}^n \Rightarrow 0 \leq y^\top A y = y^\top \lambda y = \lambda \|y\|^2$ for

---

[1] We write the transpose of matrix A as $A^\top$.

J. Suzuki, *Kernel Methods for Machine Learning with Math and R*,
https://doi.org/10.1007/978-981-19-0398-4_1

an eigenvalue λ of A and its eigenvector $y \in \mathbb{R}^n$. $3. \Rightarrow 1.$ holds since $\lambda_1, \ldots, \lambda_n \geq 0 \Rightarrow A = PDP^\top = P\sqrt{D}\sqrt{D}P^\top = (\sqrt{D}P^\top)^\top \sqrt{D}P^\top$, where D and $\sqrt{D}$ are diagonal matrices with elements $\lambda_1, \ldots, \lambda_n$ and $\sqrt{\lambda_1}, \ldots, \sqrt{\lambda_n}$, and P is the corresponding orthogonal matrix.   □

A nonnegative definite matrix A is symmetric. In this book, we say that a nonnegative definite matrix is positive definite if all of its eigenvalues are positive. In addition, we assume that the elements of any matrix are real. However, the following fact is often useful when we deal with complex numbers and Fourier transformations.

Corollary 1 *For a nonnegative definite matrix $A \in \mathbb{R}^{n \times n}$, we have that $z^\top A\bar{z} \geq 0$ for any $z \in \mathbb{C}^n$, where $i = \sqrt{-1}$ is the imaginary unit, and we write the conjugate $x - iy$ of $z = x + iy \in \mathbb{C}$ with $x, y \in \mathbb{R}$ as $\bar{z}$.*

Proof. Since there exists a $B \in \mathbb{R}^{n \times n}$ such that $A = B^\top B$ for a nonnegative definite matrix $A \in \mathbb{R}^{n \times n}$, we have that

$$z^\top A\bar{z} = z^\top B^\top B\bar{z} = (Bz)^\top \overline{Bz} = |Bz|^2 \geq 0$$

for any $z = [z_1, \ldots, z_n] \in \mathbb{C}^n$.   □

Example 1

```
n=3
B=matrix(rnorm(n^2),3,3)
A=t(B)%*%B
eigen(A)
```

```
eigen() decomposition
$values
[1] 4.39110234 0.30991246 0.07614846
$vectors
            [,1]          [,2]        [,3]
[1,] -0.5240328   0.83123427 0.1855780
[2,] -0.8043386  -0.55464891 0.2130822
[3,]  0.2800519  -0.03760552 0.9592480
```

```
S=NULL
for(i in 1:10){
  z=rnorm(n)
  y=drop(t(z)%*%A%*%z)
  S=c(S,y)
}
print(S)
```

```
[1]  0.1457017  9.4622216 21.4300930  0.9116660 14.3378729
[6]  7.1619008  7.7995278  0.5646901  5.1535156  0.9163690
```

1.2 Kernels

Let E be a set. We often express similarity between elements $x, y \in E$ by using a bivariate function $k : E \times E \to \mathbb{R}$ not just for data analysis but also for various information processing tasks. The larger $k(x, y)$ is, the more similar x, y are. We call such a function $k : E \times E \to \mathbb{R}$ a kernel.

Example 2 (*Epanechnikov kernel*) We use the kernel $k : E \times E \to \mathbb{R}$ such that

$$k(x, y) = D \left(\frac{|x - y|}{\lambda} \right)$$

$$D(t) = \begin{cases} \dfrac{3}{4}(1 - t^2), & |t| \leq 1 \\ 0, & Otherwise \end{cases}$$

for $\lambda > 0$, and we construct the following function (the Nadaraya-Watson estimator) from observations $(x_1, y_1), \ldots, (x_N, y_N) \in E \times \mathbb{R}$:

$$\hat{f}(x) = \frac{\sum_{i=1}^{N} k(x, x_i) y_i}{\sum_{j=1}^{N} k(x, x_j)} .$$

For a given input $x_* \in E$ that is different from the N pairs of inputs, we return the weighted sum of $y_1, \ldots, y_N$,

$$\frac{k(x_*, x_1)}{\sum_{j=1}^{N} k(x_*, x_j)}, \ldots, \frac{k(x_*, x_N)}{\sum_{j=1}^{N} k(x_*, x_j)},$$

as the output $\hat{f}(x_*)$. Because we assume that a larger $k(x, y)$ yields a more similar $x, y \in E$, the more similar x_* and x_i are, the larger the weight of y_i.

Given an input $x_* \in E$ for $i = 1, \ldots, N$, we weight y_i such that $x_i - \lambda \leq x_* \leq x_i + \lambda$ is proportional to $k(x_i, x_*)$. If we make the λ value smaller, we predict y_* by using only the (x_i, y_i) for which x_i and x_* are close. We display the output obtained when we execute the following code in Fig. 1.1.

```
1   n=250; x=2*rnorm(n); y=sin(2*pi*x)+rnorm(n)/4   ## Data Generation
2   D=function(t) max(0.75*(1-t^2),0)               ## Function Def D
3   k=function(x,y,lambda) D(abs(x-y)/lambda)       ## Function Def K
4   f=function(z,lambda){                           ## Function Def f
5     S=0; T=0;
6     for(i in 1:n){S=S+k(x[i],z,lambda)*y[i]; T=T+k(x[i],z,lambda)}
7     return(S/T)
8   }
9   plot(seq(-3,3,length=10),seq(-2,3,length=10),type="n",xlab="x",
         ylab="y"); points(x,y)
10  xx=seq(-3,3,0.1)
11  yy=NULL;for(zz in xx)yy=c(yy,f(zz,0.05)); lines(xx,yy,col="green")
```

```
12  yy=NULL;for(zz in xx)yy=c(yy,f(zz,0.35)); lines(xx,yy,col="blue")
13  yy=NULL;for(zz in xx)yy=c(yy,f(zz,0.50)); lines(xx,yy,col="red")
14  title("Nadaraya-Watson Estimator")
15  legend("topleft",legend=paste0("lambda=",c(0.05, 0.35, 0.50)),
16          lwd=1,col=c("green","blue","red"))
```

1.3   Positive Definite Kernels

The kernels that we consider in this book satisfy the positive definiteness criterion defined below. Suppose $k : E \times E \to \mathbb{R}$ is symmetric, i.e., $k(x, y) = k(y, x), x, y \in E$. For $x_1, \ldots, x_n \in E$ ($n \geq 1$), we say that the matrix

$$
\begin{bmatrix} k(x_1, x_1) & \cdots & k(x_1, x_n) \\ \vdots & \ddots & \vdots \\ k(x_n, x_1) & \cdots & k(x_n, x_n) \end{bmatrix} \in \mathbb{R}^{n \times n} \tag{1.1}
$$

is the Gram matrix w.r.t. a k of order n. We say that k is a positive definite kernel[2] if the Gram matrix of order n is nonnegative definite for any $n \geq 1$ and $x_1, \ldots, x_n \in E$.

Example 3 The kernel in Example 2 does not satisfy positive definiteness. In fact, when $\lambda = 2$, $n = 3$, and $x_1 = -1$, $x_2 = 0$, $x_3 = 1$, the matrix consisting of $K_\lambda(x_i, y_i)$ can be written as

$$
\begin{bmatrix} k(x_1, x_1) & k(x_1, x_2) & k(x_1, x_3) \\ k(x_2, x_1) & k(x_2, x_2) & k(x_2, x_3) \\ k(x_3, x_1) & k(x_3, x_2) & k(x_3, x_3) \end{bmatrix} = \begin{bmatrix} 3/4 & 9/16 & 0 \\ 9/16 & 3/4 & 9/16 \\ 0 & 9/16 & 3/4 \end{bmatrix}
$$

Fig. 1.1 We use the Epanechnikov kernel and Nadaraya-Watson estimator to draw the curves for $\lambda = 0.05, 0.35, 0.5$. Finally, we obtain the optimal λ value and present it in the same graph

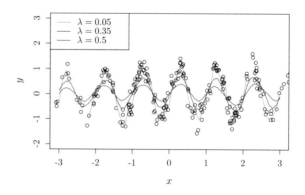

---

[2] Although it seems appropriate to say "a nonnegative definite kernel", the custom of saying "a positive definite kernel" has been established.

and the determinant is computed as $3^3/2^6 - 3^5/2^{10} - 3^5/2^{10} = -3^3/2^9$. In general, the determinant of a matrix is the product of its eigenvalues, and we find that at least one of the three eigenvalues is negative.

Example 4 For random variables $\{X_i\}_{i=1}^{\infty}$ that are not necessarily independent, if $k(X_i, X_j)$ is the covariance between X_i, X_j, the Gram matrix of any order is the covariance matrix among a finite number of X_j, which means that k is positive definite. We discuss Gaussian processes based on this fact in Chap. 6.

By assuming positive definiteness, the theory of kernels will be developed in this book. Hereafter, when we state kernels, we are referring to positive definite kernels.

Let H be a linear space (vector space) equipped with an inner product $\langle \cdot, \cdot \rangle_H$. Then, we often construct a positive definite kernel with

$$k(x, y) = \langle \Psi(x), \Psi(y) \rangle_H. \tag{1.2}$$

By using an arbitrary map $\Psi : E \to H$. We say that such a Ψ is a feature map. In this chapter, we may assume that the linear space H is the Euclidean space $H = \mathbb{R}^d$ of dimensionality d with the standard inner product $\langle x, y \rangle_{\mathbb{R}^d} = x^\top y, x, y \in \mathbb{R}^d$. We define the linear space and inner product concepts in Chap. 2.

Proposition 2 *The kernel* $k : E \times E \to \mathbb{R}$ *defined in (1.2) is positive definite.*

Proof. We arbitrarily fix $n = 1, 2, \cdots$ and $x_1, \cdots, x_n \in E$ and denote the Gram matrix (1.1) by K. Then, from the definition of inner products, for an arbitrary $z = [z_1, \cdots, z_n] \in \mathbb{R}^n$, we have

$$z^\top K z = \sum_{i=1}^{n} \sum_{j=1}^{n} z_i z_j \langle \Psi(x_i), \Psi(x_j) \rangle_H = \langle \sum_{i=1}^{n} z_i \Psi(x_i), \sum_{j=1}^{n} z_j \Psi(x_j) \rangle_H = \| \sum_{j=1}^{n} z_j \Psi(x_j) \|_H^2 \geq 0,$$

where we write $\|a\|_H := \langle a, a \rangle_H^{1/2}$ for $a \in H$. □

Proposition 3 *If the matrices* A, B *are nonnegative definite, then so is the Hadamard product*
$$A \circ B$$
(elementwise multiplication).

Proof. See the appendix at the end of this chapter.

Proposition 3 is helpful for proving the second part of the following proposition.

Proposition 4 *If the kernels* $k_1, k_2, \ldots$ *are positive definite, then so are the following* $E \times E \to \mathbb{R}$:

1. $ak_1 + bk_2$ $(a, b \geq 0)$,
2. $k_1 k_2$,

3. the limit[3] of $\{k_i\}$ when it converges,
4. k has only one value $a \geq 0$ (constant function), and
5. $f(x)k(x, y)f(y)$ $(x, y \in E)$ for an arbitrary $f : E \to \mathbb{R}$,

where the third point claims that the limit $k_\infty(x, y) := \lim_{i \to \infty} k_i(x, y)$ satisfies positive definiteness for any $x, y \in E$.

Proof. $ak_1 + bk_2$ is positive definite because

$$x^\top Ax \geq 0, x^\top Bx \geq 0 \Rightarrow x^\top (aA + bB)x \geq 0$$

for $A, B \in \mathbb{R}^{n \times n}$. The product $k_1 k_2$ is positive definite because if $A = (A_{i,j})$, $B = (B_{i,j})$ are nonnegative definite, then so is the Hadamard product $A \circ B$ (Proposition 3). The third statement assumes the existence of a positive integer n such that

$$B_\infty = \sum_{j=1}^n \sum_{h=1}^n z_j z_h k_\infty(x_j, x_h) = -\epsilon$$

for $x_1, \cdots, x_n \in E$, $z_1, \ldots, z_n \in \mathbb{R}$, and $\epsilon > 0$. Then, the difference between $B_i :=$ $\sum_{j=1}^n \sum_{h=1}^n z_j z_h k_i(x_j, x_h) \geq 0$ and B_∞ becomes arbitrarily close to zero as $i \to \infty$. However, the difference is at least $\epsilon > 0$, which is a contradiction and means that $B_\infty \geq 0$. If a kernel takes only a (nonnegative) constant value a, since all the values in (1.1) are $a \geq 0$, we have

$$\begin{bmatrix} a & \cdots & a \\ \vdots & \ddots & \vdots \\ a & \cdots & a \end{bmatrix} = \begin{bmatrix} \sqrt{a/n} & \cdots & \sqrt{a/n} \\ \vdots & \ddots & \vdots \\ \sqrt{a/n} & \cdots & \sqrt{a/n} \end{bmatrix}^\top \begin{bmatrix} \sqrt{a/n} & \cdots & \sqrt{a/n} \\ \vdots & \ddots & \vdots \\ \sqrt{a/n} & \cdots & \sqrt{a/n} \end{bmatrix}.$$

The last claim is due to the implication

$$x^\top Ax \geq 0, \ x \in \mathbb{R}^n \Rightarrow x^\top DADx \geq 0, \ x \in \mathbb{R}^n,$$

which we can examine by substituting $y = Dx$ into $y^\top Ay \geq 0$. In particular, we may regard A and D as the matrix (1.1) and diagonal matrix with the elements $f(x_1), \cdots, f(x_n)$, respectively. □

In addition, the $f(x)f(y)$ obtained by substituting $k(x, y) = 1$ for $x, y \in E$ in the last item of Proposition 4 is positive definite. Moreover, the

$$\frac{k(x, y)}{\sqrt{k(x, x)k(y, y)}} \tag{1.3}$$

obtained by substituting $f(x) = \{k(x, x)\}^{-1/2}$ for $k(x, x) > 0$ $(x \in E)$ in the last item of Proposition 4 is positive definite. Furthermore,, the value obtained by substi-

---

[3] the limit of $k_i(x, y)$ for each $(x, y) \in E$.

tuting $n = 2$, $x_1 = x$, and $x_2 = y$ into (1.1) is nonnegative, and the absolute value of (1.3) does not exceed one. We say that (1.3) is the positive definite kernel obtained by normalizing $k(x, y)$.

Example 5 (*Linear Kernel*) Let $E := \mathbb{R}^d$. Then, the kernel $k(x, y) = x^{\top} A y = \langle Bx, By \rangle_H$, $x, y \in \mathbb{R}^d$ using the nonnegative definite matrix $A = B^{\top} B \in \mathbb{R}^{d \times d}$, $B \in \mathbb{R}^{d \times d}$ is positive definite because it corresponds to the case in which the map Ψ in Proposition 2 is $E \ni x \mapsto Bx \in H$. In particular, if A is the unit matrix, then the map Ψ is the identity map. In this sense, the positive definite kernel is an extension of the inner product $k(x, y) = x^{\top} y$.

Example 6 (*Exponential Type*) Let $\beta > 0$, $n \geq 0$, and $x, y \in \mathbb{R}^d$. Then,

$$k_m(x, y) := 1 + \beta x^{\top} y + \frac{\beta^2}{2}(x^{\top} y)^2 + \cdots + \frac{\beta^m}{m!}(x^{\top} y)^m \qquad (1.4)$$

($m \geq 1$) is a polynomial of the products of positive definite kernels, and the coefficients are nonnegative. From the first two items of Proposition 4, this kernel is a positive definite kernel. Additionally, because (1.4) is a Taylor expansion up to the order m, from the third item of Proposition 4,

$$k_\infty(x, y) := \exp(\beta x^{\top} y) = \lim_{m \to \infty} k_m(x, y)$$

is a positive definite kernel as well.

Example 7 (*Gaussian Kernel*) The kernel

$$k(x, y) := \exp\{-\frac{1}{2\sigma^2} \|x - y\|^2\}, \ \sigma > 0 \qquad (1.5)$$

for $x, y \in \mathbb{R}^d$ can be written as

$$\exp\{-\frac{1}{2\sigma^2} \|x - y\|^2\} = \exp\{-\frac{\|x\|^2}{2\sigma^2}\} \exp\{\frac{x^{\top} y}{\sigma^2}\} \exp\{-\frac{\|y\|^2}{2\sigma^2}\} .$$

Thus, from the fifth item of Proposition 4 and the fact that $\exp(\beta x^{\top} y)$ with $\beta = \sigma^{-2}$ is positive definite, we see that (7) is positive definite.

Example 8 (*Polynomial Kernel*) The kernel

$$k_{m,d}(x, y) := (x^{\top} y + 1)^m , \qquad (1.6)$$

for $x, y \in \mathbb{R}^d$, $d = 1, 2, \ldots$ is a polynomial of positive definite kernels (linear kernels $x^{\top} y$), and its coefficients are nonnegative. From the first two items of Proposition 4, (1.6) is positive definite.

Example 9 If we normalize the linear kernel by (1.3), we obtain $x^\top y/\|x\|\|y\|$, where we denote $\|a\| := \langle a, a \rangle^{1/2}$ for $a \in \mathbb{R}^n$. The Gaussian kernel (1.5) remains the same even if we normalize it. The polynomial kernel becomes

$$\left(\frac{x^\top y + 1}{\sqrt{x^\top x + 1}\sqrt{y^\top y + 1}} \right)^m$$

if we normalize it.

The converse is true for Proposition 2, which will be proven in Chap. 3: for any nonnegative definite kernel k, there exists a feature map $\Psi : E \to H$ such that $k(x, y) = \langle \Psi(x), \Psi(y) \rangle_H$.

Example 10 (*Polynomial Kernel*) Let $m, d \geq 1$. The feature map of the kernel $k_{m,d}(x, y) = (x^\top y + 1)^m$ with $x, y \in \mathbb{R}^d$ is

$$\Psi_{m,d}(x_1, \cdots, x_d) = \left(\sqrt{\frac{m!}{m_0! m_1! \cdots m_d!}} x_1^{m_1} \cdots x_d^{m_d} \right)_{m_0, m_1, \ldots, m_d \geq 0},$$

where the indices $(m_0, m_1, \cdots, m_d)$ range over $m_0, m_1, \cdots, m_d \geq 0$ and $m_0 + m_1 + \cdots + m_d = m$, and we assume that an order exists among the indices $(m_0, m_1, \cdots, m_d)$. If we use the multinomial theorem,

$$\left(\sum_{i=0}^{d} z_i \right)^m = \sum_{m_0 + m_1 + \cdots + m_d = m} \frac{m!}{m_0! m_1! \cdots m_d!} z_1^{m_1} \cdots z_d^{m_d}$$

($z_0 = 1$), we see that

$$(x^\top y + 1)^m = \langle \Psi_{m,d}(x), \Psi_{m,d}(y) \rangle_H$$

with $x_0 = y_0 = 1$. For example, we have

$$\Psi_{1,2}(x_1, x_2) = [1, x_1, x_2]$$

$$\Psi_{2,2}(x_1, x_2) = [1, x_1^2, x_2^2, \sqrt{2}x_1, \sqrt{2}x_2, \sqrt{2}x_1 x_2]$$

because

$$\langle \Psi_{2,1}(x_1, x_2), \Psi_{2,1}(y_1.y_2) \rangle_H = 1 + x_1 y_1 + x_2 y_2 = 1 + x^\top y = k(x, y)$$

$$\langle \Psi_{2,2}(x_1, x_2), \Psi_{2,2}(y_2.y_2) \rangle_H = 1 + x_1^2 y_1^2 + x_2^2 y_2^2 + 2x_1 y_1 + 2x_2 y_2 + 2x_1 x_2 y_1 y_2$$
$$= (1 + x_1 y_1 + x_2 y_2)^2 = (1 + x^\top y)^2 = k(x, y) .$$

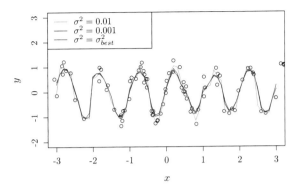

Fig. 1.2 Smoothing by predicting the values of x outside the N sample points via the Nadaraya-Watson estimator. We choose the best parameter for the Gaussian kernel via cross validation

	Group 1	Group 2	$\cdots$	Group $k-1$	Group k
First	Test	Estimation	$\cdots$	Estimation	Estimation
Second	Estimation	Test	$\cdots$	Estimation	Estimation
$\vdots$	$\vdots$	$\vdots$	$\ddots$	$\vdots$	$\vdots$
$(k-1)$-th	Estimation	Estimation	$\cdots$	Test	Estimation
k-th	Estimation	Estimation	$\cdots$	Estimation	Test

Fig. 1.3 A rotation employed for cross validation. Each group consists of N/k samples; we divide the samples into k groups based on their sample IDs. $1 \sim \dfrac{N}{k}, \dfrac{N}{k} + 1 \sim \dfrac{2N}{k}, \ldots, (k-2)\dfrac{N}{k} + 1 \sim (k-1)\dfrac{N}{k}, (k-1)\dfrac{N}{k} + 1 \sim N$

Example 11 (*Infinite-Dimensional Polynomial Kernel*) Let $0 < r \le \infty, d \ge 1$, and $E := \{x \in \mathbb{R}^d \mid \|x\|_2 < \sqrt{r}\}$. Let $f : (-r, r) \to \mathbb{R}$ be C^∞. We assume that the function can be Taylor-expanded by

$$f(x) = \sum_{n=0}^{\infty} a_n x^n , \quad x \in (-r, r) .$$

If $a_0 > 0, a_1, a_2, \ldots \ge 0$, then $f(x^\top y)$ is a positive definite kernel for $x, y \in E$. The exponential type is an infinite-dimensional polynomial kernel and is positive definite.

Example 12 In Example 2, we use the Nadaraya-Watson estimator to determine the Gaussian kernel.

```
K=function(x,y,sigma2)exp(-norm(x-y,"2")^2/2/sigma2)
f=function(z,sigma2){ ## Function Deff
  S=0; T=0;
  for(i in 1:n){S=S+K(x[i],z,sigma2)*y[i]; T=T+K(x[i],z,sigma2)}
  return(S/T)
}
```

We often obtain the optimal value for each of the kernel parameters via cross validation (CV).[4] If the parameters take continuous values, we select a finite number of candidates and obtain the evaluation value for each parameter as follows. Divide the N samples into K groups and conduct estimation with the samples belonging to group $K - 1$ group. Perform testing with the samples belonging to the one remaining group and calculate the corresponding score. Repeat the procedure K times (changing the test group) and find the sum of the obtained scores. In that way, we evaluate the performance of the kernel based on one parameter. Execute this process for all parameter candidates and use the parameters with the best evaluation values Figs. (1.2 and 1.3).

We obtain the optimal value of the parameter σ^2 via CV. We execute this procedure, setting $\sigma^2 = 0.01, 0.001$.

```
1  n=100; x=2*rnorm(n); y=sin(2*pi*x)+rnorm(n)/4 ## Data Generation
2  plot(seq(-3,3,length=10),seq(-2,3,length=10),type="n",xlab="x",
       ylab="y"); points(x,y)
3  xx=seq(-3,3,0.1)
4  yy=NULL;for(zz in xx)yy=c(yy,f(zz,0.001)); lines(xx,yy,col="green")
5  yy=NULL;for(zz in xx)yy=c(yy,f(zz,0.01)); lines(xx,yy,col="blue")
```

```
1  ## Thus far, the curves with sigma2=0.01, 0.001 have been shown
2  m=n/10
3  sigma2.seq=seq(0.001,0.01,0.001); SS.min=Inf
4  for(sigma2 in sigma2.seq){
5      SS=0
6      for(k in 1:10){
7          test=((k-1)*m+1):(k*m); train=setdiff(1:n,test)
8          for(j in test){
9              u=0; v=0;
10             for(i in train){
11                 kk=K(x[i],x[j],sigma2); u=u+kk*y[i]; v=v+kk ##
       Use of the kernel
12             }
13             if(v!=0){z=u/v; SS=SS+(y[j]-z)^2}
14         }
15     }
16     if(SS<SS.min){SS.min=SS;sigma2.best=sigma2}
17 }
18 paste0("Best sigma2 = ",sigma2.best)
19 ## Thus far, the optimum lambda has been computed
```

```
1  yy=NULL;for(zz in xx)yy=c(yy,f(zz,sigma2.best)); lines(xx,yy,col="
       red")
2  title("Nadaraya-Watson Estimator")
3  legend("topleft",legend=paste0("sigma2=",c(0.01, 0.001, "sigma2.best
       ")),
4  lwd=1,col=c("green","blue","red"))
```

---

[4] Joe Suzuki, "Statistical Learning with Math and R", Chap. 4, Springer.

1.4 Probability

Each set is an event when the sets are closed by set operations (union, intersection, and complement).

Example 13 We consider a set consisting of the subsets of $E = \{1, 2, 3, 4, 5, 6\}$ (dice eyes) that are closed by set operations:

$$\{E, \{\}, \{1, 3\}, \{5\}, \{2, 4, 6\}, \{1, 3, 5\}, \{2, 4, 5, 6\}, \{1, 2, 3, 4, 6\}\} .$$

If any of these eight elements undergo the union, intersection, or complement operations, the result remains one of these eight elements. In that sense, we can say that these eight elements are closed by the set operations. The subsets $\{1, 3\}$ and $\{2, 4, 5, 6\}$ are events, but $\{2, 4\}$ is not. On the other hand, for the entire set E, if we include $\{1\}, \{2\}, \{3\}, \{4\}, \{5\}, \{6\}$ as events, 2^6 events should be considered. Even if the entire set E is identical, whether it is an event differs depending on the set $\mathcal{F}$ of events.

In the following, we start our discussion after defining the entire set E and the set $\mathcal{F}$ of subsets (events) of E closed by the set operations. Any open interval (a, b) with $a, b \in \mathbb{R}$ is a subset of the whole real number system $\mathbb{R}$. Applying set operations (union, set product, and set complement) to multiple open intervals does not form an open interval, but the result remains a subset of $\mathbb{R}$. We call any subset of $\mathbb{R}$ obtained from an open set by set operations a Borel set of $\mathbb{R}$, and we denote such a subset as $\mathbb{B}$. A set obtained by further applying set operations to Borel sets remains a Borel set.

Example 14 For $a, b \in \mathbb{R}$, the following are Borel sets: $\{a\} = \cap_{n=1}^{\infty}(a - 1/n, a + 1/n)$, $[a, b) = \{a\} \cup (a, b)$, $(a, b] = \{b\} \cup (a, b)$, $[a, b] = \{a\} \cup (a, b]$, $\mathbb{R} = \cup_{n=0}^{\infty}(-2^n, 2^n)$, $\mathbb{Z} = \cup_{n=0}^{\infty}\{-n, n\}$, $[\sqrt{2}, 3) \cup \mathbb{Z}$.

As described above, we assume that we have defined the entire set E and the set $\mathcal{F}$ of events. At this time, the $\mu : \mathcal{F} \to [0, 1]$ that satisfies the following three conditions is called a probability:

1. $\mu(A) \geq 0$, $A \in \mathcal{F}$.
2. $A_i \cap A_j = \{\} \Rightarrow \mu(\cup_{i=1}^{\infty} A_i) = \sum_{i=1}^{\infty} \mu(A_i)$
3. $\mu(E) = 1$.

We say that μ is a measure if μ satisfies the first two conditions, and we say that this measure is finite if $\mu(E)$ takes a finite value. We say that $(E, \mathcal{F}, \mu)$ is either a probability space or a measure space, depending on whether $\mu(E) = 1$ or not.

For probability and measure spaces, if $\{e \in E | X(e) \in B\}$ is an event for any Borel set B, which means that $\{e \in E | X(e) \in B\} \in \mathcal{F}$, we say that the function $X : E \to \mathbb{R}$ is measurable in X. In particular, if we have a probability space, X is a random variable. Whether X is measurable depends on $(E, \mathcal{F})$ rather than $(E, \mathcal{F}, \mu)$.

The notion of measurability might be complex for a beginner to understand. However, it seems smoother if we intuitively understand that the function $X : E \to \mathbb{R}$ depends on an element of $\mathcal{F}$ rather than an element of E.

Example 15 (*Dice Eyes*) Suppose that $X : E \to \mathbb{R}$ for $E = \{1, 2, 3, 4, 5, 6\}$ is given by

$$X(e) = \begin{cases} 1, & e = 1, 3, 5 \\ 0, & e = 2, 4, 6 \end{cases}.$$

Then, if $\mathcal{F} = \{\{1, 3, 5\}, \{2, 4, 6\}, \{\}, E\}$, then X is a random variable. In fact, since X is measurable,

$$\{e \in E | X(e) \in \{1\}\} = \{1, 3, 5\},$$

$$\{e \in E | X(e) \in [-2, 3)\} = E,$$

$$\{e \in E | X(e) \in [0, 1)\} = \{2, 4, 6\}$$

for the Borel set $B = \{1\}, [-2, 3), [0, 1)$. Even if we choose the Borel set B, the set $\{e \in E | X(e) \in B\}$ is one of $\{1, 3, 5\}, \{2, 4, 6\}, \{\}, E$. On the other hand, if $\mathcal{F} = \{\{1, 2, 3\}, \{4, 5, 6\}, \{\}, E\}$, then X is not a random variable.

In the following, assuming that the function $f : E \to \mathbb{R}$ is measurable, we define the Lebesgue integral $\int_E f d\mu$. We first assume that f is nonnegative. For a sequence of exclusive subsets $\{B_k\}$ of $\mathcal{F}$, we define

$$\sum_k \left(\inf_{e \in B_k} f(e) \right) \mu(B_k) . \tag{1.7}$$

If $\cup_k B_k = E$ and the supremum of (1.7) w.r.t. $\{B_k\}$,

$$\sup_{\{B_k\}} \sum_k \left(\inf_{e \in B_k} f(e) \right) \mu(B_k),$$

takes a finite value, we say that the supremum is the Lebesgue integral of the measurable function f for $(E, \mathcal{B}, \mu)$, and we write $\int_E f d\mu$. When the function f is not necessarily nonnegative, we divide E into $E_+ := \{e \in E | f(e) \le 0\}$ and $E_- := \{e \in E | f(e) \ge 0\}$, and we define the above quantity for each of $f_+ := f$, $f_- := -f$. If both $\int f_+ d\mu$, $\int f_- d\mu$ take finite values, we say that $\int f d\mu := \int f_+ d\mu - \int f_- d\mu$ is the Lebesgue integral of f for $(E, \mathcal{B}, \mu)$.

If X is a random variable, the associated Borel sets are the events for the probability $\mu(\cdot)$. We say that the probability of event $X \le x$ for $x \in \mathbb{R}$

$$F_X(x) := \mu([-\infty, x)) = \int_{(-\infty, x]} d\mu$$

is the distribution function and that f_X is the probability density function of X if we can write F_X as

$$F_X(x) = \int_{-\infty}^{x} f_X(t)dt \; .$$

We say that μ is absolutely continuous if the probability $\mu(B)$ diminishes when the width sum of the intervals in any Borel set B approaches zero. The necessary and sufficient condition ensuring that a probability density function exists for the probability μ is that μ is absolutely continuous. If X takes a finite number of values, the probability density function does not exist, which means that μ is not absolutely continuous. If X takes values of $a_1 < \cdots < a_m$, then the distribution function can be written as

$$F_X(x) = \sum_{j:a_j \leq x} \mu(\{a_j\}) \; .$$

Example 16 Suppose that X follows the standard Gaussian distribution. If we make $\epsilon > 0$ close to zero, $F_X(x + \epsilon) - F_X(x - \epsilon)$ (for any $x \in \mathbb{R}$) approaches zero, which means that the probability is absolutely continuous. On the other hand, suppose that X takes values of $0, 1$; even if we make $\epsilon > 0$ close to zero, $F_X(1 + \epsilon) - F_X(1 - \epsilon)$ does not approach zero, which means that the probability is not absolutely continuous.

If we use the Lebesgue integral, we can express the probability without distinguishing between discrete and continuous variables.

Example 17 For $E = \mathbb{R}$, if the probability density function f_X exists, the expectation of X can be written as $\int_E x d\mu = \int_{-\infty}^{\infty} t f_X(t)dt$. On the other hand, if X takes values of $a_1 < \cdots < a_m$, we have $\int_E x d\mu = \sum_{j=1}^{m} a_j \mu(\{a_j\})$.

1.5 Bochner's Theorem

We consider the case in which a kernel is a function of the difference between $x, y \in E$. According to Bochner's theorem, which is the main topic of this section and will be used in later chapters, the kernel should coincide with a characteristic function in terms of probability and statistics (up to a constant).

When utilizing a univariate function $\phi : E \to \mathbb{R}$, we often use kernels in the form $k(x, y) = \phi(x - y)$, such as the Gaussian kernel. The kernel k being positive definite is equivalent to the inequality

$$\sum_{i=1}^{n} \sum_{j=1}^{n} z_i z_j \phi(x_i - x_j) \geq 0 \; , \; z = [z_1, \ldots, z_n] \in \mathbb{R}^n \tag{1.8}$$

for an arbitrary $n \geq 1, x_1, \ldots, x_n \in E$.

Let $i = \sqrt{-1}$ be the imaginary unit. We define the characteristic function of a random variable X by $\varphi : \mathbb{R}^d \to \mathbb{C}$:

$$\varphi(t) := \mathbb{E}[\exp(it^\top X)] = \int_E \exp(it^\top x) d\mu(x) , \quad t \in \mathbb{R}^d ,$$

where $\mathbb{E}[\cdot]$ denotes the expectation. If μ is absolutely continuous (i.e., the probability density function f_X exists), then $\varphi(t) := \mathbb{E}[\exp(it^\top X)] = \int_E \exp(it^\top x) f_X(x) dx$ is the Fourier transformation of $f_X(x) = \frac{d\mu(x)}{dx}$, and $f_X(x)$ can be recovered from $\varphi(x)$ via the inverse Fourier transformation

$$f_X(x) = \frac{1}{2\pi} \int_{-\infty}^{\infty} \varphi(t) e^{-it^\top x} dt .$$

Example 18 The characteristic function of the Gaussian distribution with a mean of μ and a variance of σ^2, $f(x) = \frac{1}{\sqrt{2\pi}} \exp\{-\frac{(x-\mu)^2}{2\sigma^2}\}$, is

$$\begin{aligned}
\varphi(t) &= \frac{1}{\sqrt{2\pi}} \int_{-\infty}^{\infty} \exp\{itx\} \exp\{-\frac{(x-\mu)^2}{2\sigma^2}\} dx \\
&= \frac{1}{\sqrt{2\pi}} \int_{-\infty}^{\infty} \exp[-\frac{\{x - (\mu + it\sigma^2)\}^2}{2\sigma^2}] dx \cdot \exp\{i\mu t - \frac{t^2\sigma^2}{2}\} \\
&= \exp\{i\mu t - \frac{t^2\sigma^2}{2}\} .
\end{aligned}$$

The characteristic function of the Laplace distribution $f(x) = \frac{\alpha}{2} \exp\{-\alpha|x|\}$ with a parameter $\alpha > 0$ is

$$\begin{aligned}
\int_{-\infty}^{\infty} \exp\{itx\} \frac{\alpha}{2} \exp\{-\alpha|x|\} dx &= \frac{\alpha}{2} \{\int_{-\infty}^{0} \exp[(it+\alpha)x] dx + \int_{0}^{\infty} \exp[(it-\alpha)x] dx\} \\
&= \frac{\alpha}{2} \{\left[\frac{e^{(it+\alpha)x}}{it+\alpha}\right]_{-\infty}^{0} - \left[\frac{e^{(it-\alpha)x}}{it-\alpha}\right]_{0}^{\infty}\} = \frac{\alpha^2}{t^2+\alpha^2} .
\end{aligned}$$

Proposition 5 (Bochner) *Suppose that $\phi : \mathbb{R}^n \to \mathbb{R}$ is continuous. Then, Condition (1.8) holds for an arbitrary $n \geq 1$ with $x = [x_1, \ldots, x_n] \in \mathbb{R}^n$ and $z = [z_1, \ldots, z_n] \in \mathbb{R}^n$ if and only if ϕ coincides with the characteristic function w.r.t. a probability μ up to a constant, i.e., there exists a finite measure η such that*

$$\phi(t) = \int_E \exp(it^\top x) d\eta(x) , \quad t \in \mathbb{R}^n . \tag{1.9}$$

Proof: See the Appendix at the end of this chapter.

Because a kernel evaluates a similarity between two elements in E, we do not care much about the multiplication of constants. In the following, we say that a probability μ is the probability of kernel k if μ is the finite measure η when dividing the kernel k in Proposition 5 by a constant. Note that we only consider a kernel $k(\cdot, \cdot)$ whose range is real in this book, although the range of the characteristic function is generally $\mathbb{C}^n$.

In the following, we denote $\|t\|_2$ by $\sqrt{\sum_{j=1}^d t_j^2}$ for $t = [t_1, \ldots, t_d] \in \mathbb{R}^d$.

Example 19 (*Gaussian Kernel*) $k(x, y) = \exp\{-\frac{1}{2\sigma^2}\|x - y\|_2^2\}$, $x, y \in \mathbb{R}^d$ coincides with the characteristic function $\exp\{-\frac{\|t\|_2^2}{2\sigma^2}\}$, $t = x - y \in \mathbb{R}^d$ of the Gaussian distribution with a mean of 0 and a covariance matrix $(\sigma^2)^{-1} I \in \mathbb{R}^{d \times d}$.

Example 20 (*Laplacian Kernel*) $k(x, y) = \dfrac{1}{2\pi} \dfrac{1}{\|x - y\|_2^2 + \beta^2}$, $x, y \in \mathbb{R}^n$ coincides with the characteristic function $\dfrac{\beta^2}{\|t\|_2^2 + \beta^2}$, $t = x - y \in \mathbb{R}^n$ of the Laplace distribution with a parameter $\alpha = \beta > 0$ up to the constant multiplication $[2\pi\beta^2]^{-1}$.

We can construct the kernel for this distribution if the probability density function exists. However, if we restrict our search to the kernels whose ranges real, we need to choose the parameters so that the characteristic function takes real values. For example, the Gaussian kernel obtained by setting the mean to zero takes real values.

1.6 Kernels for Strings, Trees, and Graphs

As discussed in Chap. 4, the space E of the covariates is projected via the feature map $\Psi : E \to H$. The method of evaluating similarity via the inner product (kernel) in another linear space (RKHS) has been widely used in machine learning and data science. If the similarities between the elements of the set E are accurately represented, then this approach yields improved regression and classification processing performance. As this is a kernel configuration method, we provide the notions of convolutional and marginalized kernels and illustrate them by introducing string, tree, and graph kernels.

First, we define positive definite kernels $k_1, \ldots, k_d$ for the sets $E_1, \ldots, E_d$. Suppose that we define a set E and a map $R : E_1 \times \cdots \times E_d \to E$. Then, we define the kernel $E \times E \ni (x, y) \mapsto k(x, y) \in \mathbb{R}$ by

$$k(x, y) = \sum_{R^{-1}(x)} \sum_{R^{-1}(y)} \prod_{i=1}^d k_i(x_i, y_i), \tag{1.10}$$

where $\sum_{R^{-1}(x)}$ is the sum over $(x_1, \ldots, x_d) \in E_1 \times \cdots E_d$ such that $R(x_1, \ldots, x_d) = x$. A kernel in the form of (1.10) is called a convolutional kernel [13]. Since each $k_i(x_i, y_i)$ is positive definite, $k(x, y)$ is also positive definite (according to the first two items of Proposition 4).

Example 21 (*String Kernel*) Let Σ^p be a set of strings consisting of $p \geq 0$ characters in a finite set Σ, and let $\Sigma^* := \cup_i \Sigma^i$. For example, if $\Sigma = \{A, T, G, C\}$, we have $AGGCGTG \in \Sigma^7$. Then, we define the kernel

$$k(x, y) := \sum_{u \in \Sigma^p} c_u(x) c_u(y)$$

for $x, y \in \Sigma^*$, where $c_u(x)$ denotes the number of occurrences of $u \in \Sigma^p$ in $x \in \Sigma^*$. The following represents sample code for defining this string kernel.

```
1  string.kernel=function(x,y,p){
2    m=nchar(x)
3    n=nchar(y)
4    S=0
5    for(i in 1:m)for(j in 1:n)
6      if(substring(x,i,i+p)==substring(y,j,j+p))S=S+1
7    return(S)
8  }
```

Then, we execute the procedure.

```
1  C=c("a","b","c")
2  m=10; w=sample(C,m,rep=TRUE)
3  x=NULL; for(i in 1:m)x=paste0(x,w[i])
4  n=12; w=sample(C,n,rep=TRUE)
5  y=NULL; for(i in 1:m)y=paste0(y,w[i])
```

```
1  x
```

```
[1] "bacacbcaaa"
```

```
1  y
```

```
[1] "bbbbcacbab"
```

```
1  string.kernel(x,y,2)
```

```
[1] 3
```

Suppose that $d = 3$, $E_1 = E_3 = \Sigma^*$, and $E_2 = \Sigma^p$. Then, if we concatenate $(x_1, x_2, x_3) \in E_1 \times E_2 \times E_3$, then we may state that $R(x_1, x_2, x_3) = x \in E$. If $x_2 = u$ and $y_2 = u$ appear $c_u(x)$ times in x and $c_u(y)$ times in y, respectively, then by setting $k_1(x_1, y_1) = k_3(x_3, y_3) = 1$ and $k_2(x_2, y_2) = I(x_2 = y_2 = u)$, we have

$$c_u(x)c_u(y) = \sum_{R(x_1,x_2,x_3)=x} \sum_{R(y_1,y_2,y_3)=y} 1 \cdot I(x_2 = y_2 = u) \cdot 1$$

$$k(x, y) = \sum_u c_u(x)c_u(y) = \sum_{R(x_1,x_2,x_3)=x} \sum_{R(y_1,y_2,y_3)=y} 1 \cdot I(x_2 = y_2) \cdot 1 .$$

Thus, we observe that the string kernel can be expressed by (1.10), where $I(A)$ takes values of one and zero depending on whether condition A is satisfied.

Example 22 (*Tree Kernel*) Suppose that we assign a label to each vertex of trees x, y. We wish to evaluate the similarity between x, y based on how many subtrees are shared. We denote by $c_t(x), c_t(y)$ the numbers of occurrences of subtree t in x, y, respectively. Then, the kernel

$$k(x, y) := \sum_t c_t(x)c_t(y) \tag{1.11}$$

is positive definite. In fact, for $x_1, \ldots, x_n \in E$ and an arbitrary $z_1, \ldots, z_n \in \mathbb{R}$, we have

$$\sum_{i=1}^n \sum_{j=1}^n z_i z_j k(x_i, x_j) = \sum_t \{\sum_{i=1}^n z_i c_t(x_i)\}^2 \geq 0 .$$

Let V_x, V_y be the sets of vertices in trees x, y, respectively; we write $I(u, t) = 1$ and $I(u, t) = 0$ depending on whether t has u as a vertex or not. Since (1.11) can be written as $c_t(x) = \sum_{u \in V_x} I(u, t), c_t(y) = \sum_{v \in V_y} I(v, t)$, we have

$$k(x, y) = \sum_{u \in V_x} \sum_{v \in V_y} \sum_t I(u, t)I(v, t) = \sum_{u \in V_x} \sum_{v \in V_y} c(u, v) ,$$

where $c(u, v) = \sum_t I(u, t)I(v, t)$ is the number of common subtrees in x and y such that the vertices $u \in V_x$ and $v \in V_y$ are their roots. We assume that a label $l(v)$ is assigned to each $v \in V$ and determine whether they coincide.

1. For the descendants $u_1, \ldots, u_m$ and $v_1, \ldots, v_n$ of u and v, if any of the following hold, then we define $c(u, v) := 0$:

   (a) $l(u) \neq l(v)$;
   (b) $m \neq n$;
   (c) there exists $i = 1, \ldots, m$ such that $l(u_i) \neq l(v_i)$.

2. Otherwise, we define

$$c(u, v) := \prod_{i=1}^m \{1 + c(u_i, v_i)\}.$$

For example, suppose that we assign one of the labels A, T, G, C to each vertex in Fig. 1.4. We may write this in an R function as follows, where we assume that we assign no identical labels to the vertices at the same level of the tree. Note that the function calls itself (it is a recursive function). For example, the function requires the value $C(4, 2)$ when it obtains $C(1, 1)$.

```
1   C=function(i,j){
2     S=s[[i]]; T=t[[j]]
3   ## Return 0 if the labels of the verteces i and j of trees s and t do not coincide.
4     if(S[[1]]!=T[[1]])return(0)
5   ## Return 0 if either i (of s) or j of (t) does not have any descendant.
6     if(is.null(S[[2]]))return(0)
7     if(is.null(T[[2]]))return(0)
8     if(length(S[[2]])!=length(T[[2]]))return(0)
9     U=NULL; for(x in S[[2]])U=c(U,s[[x]][[1]]); U1=sort(U)
10    V=NULL; for(y in T[[2]])V=c(V,t[[y]][[1]]); V1=sort(V)
11    m=length(U)
12  ## Return 0 if the lavels of descendants
13    for(h in 1:m)if(U1[h]!=V1[h])return(0)
14    U2=S[[2]][order(U)]
15    V2=T[[2]][order(V)]
16    W=1; for(h in 1:m)W=W*(1+C(U2[h],V2[h]))
17    return(W)
18  }
19  k=function(s,t){
20    m=length(s); n=length(t)
21    kernel=0
22    for(i in 1:m)for(j in 1:n)if(C(i,j)>0)kernel=kernel+C(i,j)
23    return(kernel)
24  }
```

```
## Describe each tree by lists with the labels and their descendants as
   vectors
> s=list()
> s[[1]]=list("G",c(2,4)); s[[2]]=list("T",3);   s[[3]]=list("C",NULL)
> s[[4]]=list("A",c(5,6)); s[[5]]=list("C",NULL); s[[6]]=list("T",NULL)
> t=list()
> t[[1]]=list("G",c(2,5)); t[[2]]=list("A",c(3,4)); t[[3]]=list("C",NULL)
> t[[4]]=list("T",NULL); t[[5]]=list("T",c(6,7)); t[[6]]=list("C",NULL)
> t[[7]]=list("A",c(8,9)); t[[8]]=list("C",NULL); t[[9]]=list("T",NULL)

> for(i in 1:6)for(j in 1:9)if(C(i,j)>0)print(c(i,j,C(i,j)))
[1] 1 1 2
[1] 4 2 1
[1] 4 7 1

> k(s,t)
[1] 4
```

Thus, the sum 4 will be the kernel value.

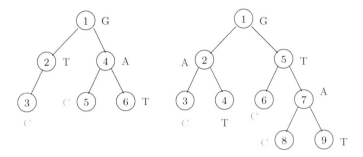

Fig. 1.4 A tree kernel evaluates the similarity in terms of which of the labels A, G, C, T are assigned to the vertices of the trees

Let X and Y be discrete random variables that take values in E_X and E_Y, respectively, and let $P(y|x)$ be the conditional probability of $X = x \in E_X$ given $Y = y \in E_Y$. Suppose that we are given a positive definite kernel $k_{XY} : E_{XY} \times E_{XY} \to \mathbb{R}$ for $E_{XY} := E_X \times E_Y$. We define the marginalized kernel by

$$k(x, x') := \sum_{y \in E_Y} \sum_{y' \in E_Y} k_{XY}((x, y), (x', y')) P(y|x) P(y'|x') , \quad x, x' \in E_X \quad (1.12)$$

for $x, x' \in E_X$ (Tsuda et al. [32]). We claim that the marginalized kernel is positive definite. In fact, k_{XY} being positive definite implies the existence of the feature map $\Psi : E_{XY} \ni (x, y) \mapsto \Psi(x, y)$ such that

$$k_{XY}((x, y), (x', y')) = \langle \Psi((x, y)), \Psi((x', y')) \rangle .$$

Thus, there exists another feature map $E_X \ni x \mapsto \sum_{y \in E_Y} P(y|x) \Psi((x, y))$ such that

$$k(x, x') := \sum_{y \in E_Y} \sum_{y' \in E_Y} P(y|x) P(y'|x') \langle \Psi((x, y)), \Psi((x', y')) \rangle$$
$$= \langle \sum_{y \in E_Y} P(y|x) \Psi((x, y)), \sum_{y' \in E_Y} P(y'|x') \Psi((x', y')) \rangle .$$

We may define (1.12) for the conditional density function f of Y given X as follows:

$$k(x, x') := \int_{y \in E_Y} \int_{y' \in E_Y} \int k_{Y|X}((x, y), (x', y')) f(y|x) f(y'|x') dy dy'$$

for $x, x' \in E_X$.

Example 23 (*Graph Kernel (Kashima et al. [19])* We construct a kernel that expresses the similarity between (directed) graphs G_1, G_2 that may contain a loop according to the set of paths connecting two vertices.

Let V, E be the sets of vertices and (directed) edges, respectively. We express each path of length m by a sequence consisting of vertices and edges: $(v_0, e_1, \ldots, e_m, v_m)$, $v_0, v_1, \ldots, v_m \in V$, and $e_1, \ldots, e_m \in E$. We assume that a label is assigned to each of the vertices and edges of the two graphs, and we define the probability of the sequence $\pi = (v_0, e_1, \ldots, e_m, v_m)$ by the products of the associated conditional probabilities $p(\pi) := p(v_0)p(v_1|v_0) \cdots p(v_m|v_{m-1})$. To this end, we consider a random walk in which we first choose $v_0 \in V$ with a probability of $p(v_0) = 1/|V|$ ($|V|$: the cardinality of V) and repeatedly choose either to stop at that point with a probability of p or to move to a neighbor vertex via one of the connected directed edges with an equiprobability times $1 - p$, where the stopping probability $0 < p < 1$ should be determined in an a priori manner. For example, if the random walk arrives at a vertex v that connects to $|V(v)|$ vertices, then the probability of moving to one of the neighboring vertices is $(1 - p)/|V(v)|$. For example, for $1 \to 4 \to 3 \to 5 \to 3$ in Fig. 1.5, the labels are $A, e, A, d, D, a, B, c, D$. If $p = 1/3$, then the probability of the directed path can be obtained via the following code:

```
k=function(s,p) prob(s,p)/length(node)
prob=function(s,p){
  if(length(node[s[1]])==0) return(0)
  if(length(s)==1) return (p)
  m=length(s)
  if(is.element(s[2],node[[s[1]]]))return((1-p)/length(node[s[1]])*
    prob(s[2:m],p))
  else return(0)
}
```

We demonstrate the execution of the code below:

```
node=list()
node[[1]]=c(2,4); node[[2]]=4; node[[3]]=c(1,5); node[[4]]=3; node[[5]]=3
> k(c(1,4,3,5,3),1/3)
[1] 0.01316872
```

Because five vertices exist, we multiply by $1/5$, choose one of the next two transitions, and so on.

$$\frac{1}{5} \cdot \left(\frac{2}{3} \cdot \frac{1}{2}\right) \cdot \left(\frac{2}{3} \cdot 1\right) \cdot \left(\frac{2}{3} \cdot \frac{1}{2}\right) \cdot \left(\frac{2}{3} \cdot 1\right) \cdot \frac{1}{3} = \frac{2^2}{5 \times 3^5}.$$

Fig. 1.5 Evaluating similarity via a graph kernel

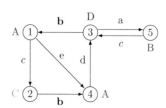

Because these probabilities are different in the directed graphs G_1, G_2, we denote them by $p(\pi|G_1)$ and $p(\pi|G)$. We express the label sequence of the path π (of length $2m + 1$) by $L(\pi)$ and define the graph kernel by

$$k(G_1, G_2) := \sum_{\pi_1} \sum_{\pi_2} p(\pi_1|G_1) p(\pi_2|G_2) I[L(\pi_1) = L(\pi_2)]$$

. We find that this kernel is a marginalized kernel if $k_{XY}((G_1, \pi_1), (G_2, \pi_2)) = I[L(\pi_1) = L(\pi_2)]$.

Appendix

Many books have proofs because Fubini's theorem, Lebesgue's dominant convergence theorem, and Levy's convergence theorem are general theorems. We have abbreviated these statements and proofs. The proof of Proposition 5 was provided by Ito [15].

Proof of Proposition 3

D is a diagonal matrix whose components are the eigenvalues $\lambda_i \geq 0$ of the non-negative definite matrix A, and U is an orthogonal matrix whose column vectors are unit eigenvectors u_i that are orthogonal to each other. Then, we can write $A = UDU^\top = \sum_{i=1}^n \lambda_i u_i u_i^\top$. Similarly, if μ_i, v_i, $i = 1, \ldots, n$ are the eigenvalues and eigenvectors of matrix B, respectively, then we can write $B = \sum_{i=1}^n \mu_i v_i v_i^\top$. At this moment, we have

$$(u_i u_i^\top) \circ (v_j v_j^\top) = (u_{i,k} u_{i,l} \cdot v_{j,k} v_{j,l})_{k,l} = (u_{i,k} v_{j,k} \cdot u_{i,l} v_{j,l})_{k,l} = (u_i \circ v_j)(u_i \circ v_j)^\top .$$

Note that this matrix is nonnegative definite. In fact, if we write $u_i \circ v_j = [y_1, \ldots, y_n] \in \mathbb{R}^n$, then component (h, l) of $(u_i \circ v_j)(u_i \circ v_j)^\top$ is $y_h y_l$, which means that $\sum_{h=1}^n \sum_{l=1}^n z_h z_l y_h y_l = (\sum_{h=1}^n z_h y_h)^2 \geq 0$ for any $z_1, \ldots, z_n$. Since matrices A and B are nonnegative definite, we have that $\lambda_i, \mu_j \geq 0$ for each $i, j = 1, \cdots, n$, which means that

$$A \circ B = \sum_{i=1}^n \sum_{j=1}^n \lambda_i \mu_j (u_i u_i^\top) \circ (v_j u_j^\top) = \sum_{i=1}^n \sum_{j=1}^n \lambda_i \mu_j (u_i \circ v_j)(u_i \circ v_j)^\top$$

is nonnegative definite. □

Proof of Proposition 5

We only show the case in which $\phi(0) = \eta(E) = 1$ because the extension is straightforward. Suppose that (1.9) holds. Then, we have

$$\sum_{j=1}^{n}\sum_{k=1}^{n} z_j z_k \phi(x_j - x_k) = \int_E \sum_{j=1}^{n} z_j e^{ix_j t} \sum_{k=1}^{n} z_k e^{-ix_k t} d\eta(t) = \int_E \left|\sum_{j=1}^{n} z_j e^{ix_j t}\right|^2 d\eta(t) \geq 0,$$

and (1.8) follows. Conversely, suppose that (1.8) holds. Since the matrix consisting of $\phi(x_i - x_j)$ for the (i, j)-th element is nonnegative definite and symmetric, we have that $\phi(x) = \phi(-x)$, $x \in \mathbb{R}$. If we substitute $n = 2$, $x_1 = u$, and $x_2 = 0$, then we obtain

$$[z_1, z_2] \begin{bmatrix} 1 & \phi(u) \\ \phi(u) & 1 \end{bmatrix} \begin{bmatrix} z_1 \\ z_2 \end{bmatrix} \geq 0$$

and $\phi(u)^2 \leq 1$ because the determinant is nonnegative. Since ϕ is bounded and continuous, it is uniformly continuous. On the other hand, $e^{-\|t\|^2/n} e^{-ixt}$ is uniformly continuous as well. In the following, we show that

$$f_n(x) := \frac{1}{2\pi} \int_{-\infty}^{\infty} \phi(t) e^{-\|t\|^2/n} e^{-ix^\top t} dt$$

is a probability density function, and the characteristic function ϕ_n approaches ϕ as $n \to \infty$. If we verify the claim, by Levy's convergence theorem [15], ϕ is the characteristic function. We show the $d = 1$ case first.

$$f_n(x) = \frac{1}{2\pi} \int_{-\infty}^{\infty} \phi(t) e^{-t^2/n} e^{-ixt} dt.$$

For $a > 0$, we have

$$\int_{-a}^{a} f_n(x) dx = \frac{1}{2\pi} \int_{-a}^{a} \int_{-\infty}^{\infty} \phi(t) e^{-t^2/n} e^{-ixt} dt dx = \frac{1}{2\pi} \int_{-\infty}^{\infty} \phi(t) e^{-t^2/n} \frac{2 \sin at}{t} dt,$$

where we use Fubini's theorem for the last equality. Then, for $b > 0$, from $\int_0^b \sin(at) da = \frac{1-\cos bt}{t} \geq 0$, $\int_{-\infty}^{\infty} \frac{1 - \cos t}{t^2} dt = \pi$, and $\phi(0) = 1$, as $b \to \infty$, we have

$$\frac{1}{b} \int_0^b \left\{ \int_{-a}^{a} f_n(x) dx \right\} da = \frac{1}{b} \int_0^b \frac{1}{2\pi} \int_{-\infty}^{\infty} \phi(t) e^{-t^2/n} \frac{2 \sin at}{t} da dt$$

$$= \frac{1}{2\pi} \int_{-\infty}^{\infty} \phi(t) e^{-t^2/n} \frac{2(1 - \cos tb)}{t^2 b} dt = \frac{1}{2\pi} \int_{-\infty}^{\infty} \phi(\frac{u}{b}) e^{-(u/b)^2/n} \frac{2(1 - \cos u)}{u^2} du \to 1,$$

where we use the dominant convergence theorem for the last equality. In general, for a $g : \mathbb{R} \to \mathbb{R}$ that is monotonically increasing and bounded from above, we have

$$\lim_{y \to \infty} \frac{1}{y} \int_0^y g(x)dx = \lim_{x \to \infty} g(x).$$

Thus, we have $\int_{-\infty}^{\infty} f_n(x)dx = 1$.

Finally, we show that $\phi_n \to \phi \ (n \to \infty)$:

$$
\begin{aligned}
\phi_n(z) &:= \lim_{a \to \infty} \int_{-a}^{a} e^{iza} \frac{1}{2\pi} \int_{-\infty}^{\infty} \phi(t)e^{-t^2/n} e^{-ita} dt \\
&= \lim_{a \to \infty} \frac{1}{2\pi} \int_{-\infty}^{\infty} \phi(t)e^{-t^2/n} \frac{2 \sin a(t-z)}{t-z} dt \\
&= \lim_{b \to \infty} \frac{1}{b} \int_0^b da \frac{1}{2\pi} \int_{-\infty}^{\infty} \phi(t)e^{-t^2/n} \frac{2 \sin a(t-z)}{t-z} dt \\
&= \lim_{b \to \infty} \frac{1}{2\pi} \int_{-\infty}^{\infty} \phi(t)e^{-t^2/n} \frac{2(1-\cos b(t-z))}{b(t-z)^2} dt \\
&= \lim_{b \to \infty} \frac{1}{2\pi} \int_{-\infty}^{\infty} \phi(z+\frac{s}{b})e^{-(z+s/b)^2/n} \frac{2(1-\cos s)}{s^2} ds = \phi(z)e^{-z^2/n} \to \phi(z).
\end{aligned}
$$

For a general $d \geq 1$, if we use $\|t\|_2^2 = t_1^2 + \ldots + t_d^2$,

$$\int_{-a_1}^{a_1} \cdots \int_{-a_d}^{a_d} e^{-i(x_1 t_1 + \cdots x_d t_d)} dx_1 \cdots dx_d = \frac{2 \sin a_1 x_1}{t_1} \cdots \frac{2 \sin a_d x_d}{t_d},$$

and

$$\int_0^{b_i} \frac{2 \sin a_i x_i}{t_i} da_i = \frac{2(1 - \cos t_i b_i)}{t_i^2 b_i},$$

$(i = 1, \ldots, d)$, then the same claim can be obtained. $\square$

Exercises 1~15

1. Show that the following three conditions are equivalent for a symmetric matrix $A \in \mathbb{R}^{n \times n}$.

   (a) There exists a square matrix B such that $A = B^\top B$;
   (b) $x^\top A x \geq 0$ for an arbitrary $x \in \mathbb{R}^n$;
   (c) All the eigenvalues of A are nonnegative.

   In addition, using the R language generate a square matrix $B \in \mathbb{R}^{n \times n}$ with real elements by generating random numbers to obtain $A = B^\top B$. Then, randomly generate five more $x \in \mathbb{R}^n \ (n = 5)$ to examine whether $x^\top A x$ is nonnegative for each value.

2. Consider the Epanechnikov kernel defined by $k : E \times E \to \mathbb{R}$

$$k(x, y) = D\left(\frac{|x - y|}{\lambda}\right)$$

$$D(t) = \begin{cases} \dfrac{3}{4}(1 - t^2), & |t| \leq 1 \\ 0, & \text{Otherwise} \end{cases}$$

for $\lambda > 0$. Suppose that we write a kernel for $\lambda > 0$ and $(x, y) \in E \times E$ in the R language as shown below:

```
1  k=function(x,y,lambda) D(abs(x-y)/lambda).
```

Specify the function D using the R language. Moreover, define the function f that makes a prediction at $z \in E$ based on the Nadaraya-Watson estimator by utilizing the function k such that z, λ are the inputs of f and k, respectively, and $(x_1, y_1), \ldots, (x_N, y_N)$ are global. Then, execute the following to examine whether the functions D, f work properly.

```
1  n=250; x=2*rnorm(n); y=sin(2*pi*x)+rnorm(n)/4
2  plot(seq(-3,3,length=10),seq(-2,3,length=10),type="n",xlab="x",
        ylab="y"); points(x,y)
3  xx=seq(-3,3,0.1)
4  yy=NULL;for(zz in xx)yy=c(yy,f(zz,0.05)); lines(xx,yy,col="green
        ")
5  yy=NULL;for(zz in xx)yy=c(yy,f(zz,0.50)); lines(xx,yy,col="red")
6  title("Nadaraya-Watson estimator")
7  legend("topleft",legend=paste0("lambda=",c(0.05, 0.35, 0.50)),
8          lwd=1,col=c("green","blue","red"))
```

Replace the Epanechnikov kernel with the Gaussian kernel, the exponential type, and the polynomial kernel and execute them.

3. Show that the determinant of $A \in \mathbb{R}^{3 \times 3}$ coincides with the product of the three eigenvalues. In addition, show that if the determinant is negative, at least one of the eigenvalues is negative.

4. Show that the Hadamard product of nonnegative definite matrices of the same size is nonnegative definite. Show also that the kernel obtained by multiplying positive definite kernels is positive definite.

5. Show that a square matrix whose elements consist of the same nonnegative value is nonnegative definite. Show further that a kernel that outputs a nonnegative constant is positive definite.

6. Find the feature map $\Psi_{3,2}(x_1, x_2)$ of the polynomial kernel $k_{3,2}(x, y) = (x^\top y + 1)^3$ for $x, y \in \mathbb{R}^2$ to derive

$$k_{3,2}(x, y) = \Psi_{3,2}(x_1, x_2)^\top \Psi_{3,2}(x_1, x_2) .$$

7. Use Proposition 4 to show that the Gaussian and polynomial kernels and exponential types are positive definite. Show also that the kernel obtained by nor-

malizing a positive definite kernel is positive definite. What kernel do we obtain when we normalize the exponential type and the Gaussian kernel?

8. The following procedure chooses the optimal parameter σ^2 of the Gaussian kernel via tenfold CV when applying the Nadaraya-Watson estimator to the samples. Change the tenfold CV procedure to the N-fold (leave-one-out) CV process to find the optimal σ^2, and draw the curve by executing the procedure below.

```
1   k=function(x,y,sigma2)exp(-(x-y)^2/2/sigma2)
2   n=100; x=2*rnorm(n); y=sin(2*pi*x)+rnorm(n)/4 ## Data Generation
3   m=n/10
4   sigma2.seq=seq(0.001,0.01,0.001); SS.min=Inf
5   for(sigma2 in sigma2.seq){
6     SS=0
7     for(h in 1:10){
8       test=((h-1)*m+1):(h*m); train=setdiff(1:n,test)
9       for(j in test){
10        u=0; v=0;
11        for(i in train){
12          kk=k(x[i],x[j],sigma2); u=u+kk*y[i]; v=v+kk
13        }
14        z=u/v
15        SS=SS+(y[j]-z)^2
16      }
17    }
18    if(SS<SS.min){SS.min=SS;sigma2.best=sigma2}
19  }
20  paste0("Best sigma2 = ",sigma2.best)
21  plot(seq(-3,3,length=10),seq(-2,3,length=10),type="n",xlab="x",
        ylab="y"); points(x,y)
22  xx=seq(-3,3,0.1)
23  yy=NULL;for(zz in xx)yy=c(yy,f(zz,sigma2.best)); lines(xx,yy,col
        ="red")
24  title("Nadaraya-Watson Estimator")
```

9. For a probability space $(E, \mathcal{F}, \mu)$ with $E = \{1, 2, 3, 4, 5, 6\}$ and a map $X : E \to \mathbb{R}$, show that if

$$X(e) = \begin{cases} 1, & e = 1, 3, 5 \\ 0, & e = 2, 4, 6 \end{cases}$$

and $\mathcal{F} = \{\{1, 2, 3\}, \{4, 5, 6\}, \{\}, E\}$, then X is not a random variable (not measurable).

10. Derive the characteristic function of the Gaussian distribution $f(x) = \frac{1}{\sqrt{2\pi}} \exp\{-\frac{(x-\mu)^2}{2\sigma^2}\}$ with a mean of μ and a variance of σ^2 and find the condition for the characteristic function to be a real function. Do the same for the Laplace distribution $f(x) = \frac{\alpha}{2} \exp\{-\alpha|x|\}$ with a parameter $\alpha > 0$.

11. Obtain the kernel value between the left tree and itself in Fig. 1.4. Construct and execute a program to find this value.

12. Randomly generate binary sequences x, y of length 10 to obtain the string kernel value $k(x, y)$.

```
1  string.kernel=function(x,y){
2    m=nchar(x)
3    n=nchar(y)
4    S=0
5    for(i in 1:m)for(j in i:m)for(k in 1:n)
6      if(substring(x,i,j)==substring(y,k,k+j−i))S=S+1
7    return(S)
8  }.
```

13. Show that the string, tree, and marginalized kernels are positive definite. Show also that the string and graph kernels are convolutional and marginalized kernels, respectively.

14. How can we compute the path probabilities below when we consider a random walk in the directed graph of Fig. 1.5 if the stopping probability is $p = 1/3$?

   (a) $3 \rightarrow 1 \rightarrow 4 \rightarrow 3 \rightarrow 5$;
   (b) $1 \rightarrow 2 \rightarrow 4 \rightarrow 1 \rightarrow 2$;
   (c) $3 \rightarrow 5 \rightarrow 3 \rightarrow 5$.

15. What inconvenience occurs when we execute the procedure below to compute a graph kernel? Illustrate this inconvenience with an example.

```
1  k=function(s,p) prob(s,p)/length(node)
2  prob=function(s,p){
3    if(length(node[s[1]])==0) return(0)
4    if(length(s)==1) return (p)
5    m=length(s)
6    S=(1−p)/length(node[s[1]])*prob(s[2:m],p)
7    return(S)
8  }.
```

Chapter 2
Hilbert Spaces

When considering machine learning and data science issues, in many cases, the calculus and linear algebra courses taken during the first year of university provide sufficient background information. However, we require knowledge of metric spaces and their completeness, as well as linear algebras with nonfinite dimensions, for kernels. If your major is not mathematics, we might have few opportunities to study these topics, and it may be challenging to learn them in a short period. This chapter aims to learn Hilbert spaces, the projection theorem, linear operators, and (some of) the compact operators necessary for understanding kernels. Unlike finite-dimensional linear spaces, ordinary Hilbert spaces require scrutiny of their completeness.

2.1 Metric Spaces and Their Completeness

Let M be a set. We say that a bivariate function $d : M \times M \to \mathbb{R}$ is a distance if

1. $d(x, y) \geq 0$;
2. $d(x, y) = 0 \iff x = y$;
3. $d(x, y) = d(y, x)$;
4. $d(x, z) \leq d(x, y) + d(y, z)$

for $x, y, z \in M$, and the pair (M, d) is a metric space[1]

Let E be a subset of the metric space M. We say that E is an open set if a positive constant ϵ exists such that $U(x, \epsilon) := \{y \in M \mid d(x, y) < \epsilon\} \subseteq E$ for each $x \in E$. Moreover, we say that $y \in M$ is a convergence point of E if $U(y, \epsilon) \cap E \neq \{\}$ for an arbitrary $\epsilon > 0$, and E is a closed set if E contains all the convergence points of E.

Example 24 The set $M = [0, 1]$ is a closed set because the neighborhood $U(y, \epsilon)$ of $y \notin M$ has no intersection with M if we make the radius $\epsilon > 0$ smaller, which means that M contains all the convergence points of M. On the other hand, $M = (0, 1)$ is

---

[1] We call M a metric space rather than (M, d) when we do not stress d or when d is apparent.

© The Author(s), under exclusive license to Springer Nature Singapore Pte Ltd. 2022
J. Suzuki, *Kernel Methods for Machine Learning with Math and R*,
https://doi.org/10.1007/978-981-19-0398-4_2

an open set because M contains the neighborhood $U(y, \epsilon)$ of $y \in M$ if we make the radius $\epsilon > 0$ smaller. If we add $\{0\}$, $\{1\}$ to the interval $(0, 1)$, $(0, 1]$, $[0, 1)$, we obtain the closed set $[0, 1]$.

We say that the minimum closed set in M that contains E is the closure of E, and we write this as $\overline{E}$. If E is not a closed set, then E does not contain all the convergence points. Thus, the closure is the set of convergence points of E. Moreover, we say that E is dense in M if $\overline{E} = M$, which is equivalent to the following conditions: "$y \in E$ exists such that $d(x, y) < \epsilon$ for an arbitrary $\epsilon > 0$ and $x \in M$", and "each point in M is a convergence point of E". Furthermore, we say that M is separable if it contains a dense subset that consists of countable points.

Example 25 For the distance $d(x, y) := |x - y|$ with $x, y \in \mathbb{R}$ and the metric space $(\mathbb{R}, d)$, each irrational number $a \in \mathbb{R}\backslash\mathbb{Q}$ is a convergence point of $\mathbb{Q}$. In fact, for an arbitrary $\epsilon > 0$, the interval $(a - \epsilon, a + \epsilon)$ contains a rational number $b \in \mathbb{Q}$. Thus, $\mathbb{Q}$ does not contain the convergence point $a \notin \mathbb{Q}$ and is not a closed set in $\mathbb{R}$. Moreover, the closure of $\mathbb{Q}$ is $\mathbb{R}$ ($\mathbb{Q}$ is dense in $\mathbb{R}$). Furthermore, since $\mathbb{Q}$ is a countable set, we find that $\mathbb{R}$ is separable.

Let (M, d) be a metric space. We say that a sequence $\{x_n\}$ in[2] M converges to $x \in M$ if $d(x_n, x) \to 0$ as $n \to \infty$ for $x \in M$, and we write this as $x_n \to x$. On the other hand, we say that a sequence $\{x_n\}$ in M is Cauchy if $d(x_m, x_n) \to 0$ as $m, n \to \infty$, i.e., if $\sup_{m,n \geq N} d(x_m, x_n) \to 0$ as $N \to \infty$.

If $\{x_n\}$ converges to some $x \in M$, then it is a Cauchy sequence. However, the converse does not hold. We say that a metric space (M, d) is complete if each Cauchy sequence $\{x_n\}$ in M converges to an element in M. We say that (M, d) is bounded if there exists a $C > 0$ such that $d(x, y) < C$ for an arbitrary $x, y \in M$, and the minimum and maximum values are the upper and lower limits if M is bounded from above and below, respectively.

Example 26 An arbitrary Cauchy sequence is bounded. In fact, for any $\epsilon > 0$, we can choose $N := N(\epsilon)$ such that $m, n \geq N \Rightarrow d(x_m, x_n) < \epsilon$, and we have

$$\min\{x_1, \ldots, x_{N-1}, x_N - \epsilon\} \leq x_n \leq \max\{x_1, \ldots, x_{N-1}, x_N + \epsilon\}.$$

Example 27 ($\mathbb{Q}$ is Not Complete) The sequence $\{a_n\}$ defined by $a_1=1$, $a_{n+1} = \frac{1}{2}a_n + \frac{1}{a_n}$ $(n \geq 1)$ is in $\mathbb{Q}$. However, we can prove that $\{a_n\}$ is a Cauchy sequence in $\mathbb{Q}$ but $a_n \to \sqrt{2} \notin \mathbb{Q}$ (Exercise 17).

Proposition 6 $\mathbb{R}$ is complete.

Proof: If $\{x_n\}$ is a Cauchy sequence in $\mathbb{R}$, then $\{x_n\}$ is bounded (Example 26). If we write the upper and lower limits of $\{x_n\}_{n=s}^{\infty}$ as l_s, m_s, respectively, the monotone sequences $\{m_s\}$, $\{l_s\}$ in $\mathbb{R}$ share the same limit. In fact, from the above assumption,

---

[2] $\{x_n\}$ with $x_n \in M$ for each n.

we can make $l_s - m_s = \sup\{|x_p - x_q| : p, q \geq s\}$ as small as possible. Thus, $\mathbb{R}$ is complete. □

If the number of dimensions is finite, we may check completeness for each dimension, and we see that $\mathbb{R}^p$ is complete for any $p \geq 1$.

Suppose that we arbitrarily set a neighborhood $U(P)$ for each $P \in M$ beforehand. We say that a set M is compact if there exist finite m and $P_1, \ldots, P_m \in M$ such that $M \subseteq \cup_{i=1}^m U(P_i)$.

Example 28 Let $M = (0, 1)$, and suppose that we define the neighborhood $U(x) := (\frac{1}{2}x, \frac{3}{2}x)$ for each $x \in M$ beforehand. Then, for any n and $x_1, \ldots, x_n \in M$, we have

$$(0, 1) \not\subseteq \cup_{i=1}^n (\frac{1}{2}x_i, \frac{3}{2}x_i) ,$$

which means that M is not compact.

Proposition 7 (Heine-Borel) *For $\mathbb{R}^p$, any bounded closed set M is compact.*

Proof: Suppose that we have set a neighborhood $U(P)$ for each $P \in M$ and that $M \subseteq \cup_{i=1}^m U(P_i)$ cannot be realized by any m and $P_1, \ldots, P_m$. If we divide the closed set (rectangular) that contains $M \subseteq \mathbb{R}^p$ into two components for each dimension, then at least one of the 2^p rectangles cannot be covered by a finite number of neighborhoods. If we repeat this procedure, then the volume of the rectangle that a finite number of neighborhoods cannot cover becomes sufficiently small for the center to converge to a $P^* \in M$; furthermore, we can cover the rectangle with $U(P^*)$, which is a contradiction. □

Let (M_1, d_1), (M_2, d_2) be metric spaces. We say that the map $f : M_1 \to M_2$ is continuous at $x \in M_1$ if for any for $\epsilon > 0$, there exists $\delta(x, \epsilon)$ such that for $y \in M_1$,

$$d_1(x, y) < \delta(x, \epsilon) \Rightarrow d_2(f(x), f(y)) < \epsilon . \tag{2.1}$$

In particular, if there exists $\delta(x, \epsilon)$ that does not depend on $x \in M_1$ in (2.1), we say that f is uniformly continuous.

Example 29 The function $f(x) = 1/x$ defined on the interval $(0, 1]$ is continuous but is not uniformly continuous. In fact, if we make x approach y after fixing y, we can make $d_2(f(x), f(y)) = |\frac{1}{x} - \frac{1}{y}|$ as small as possible, which means that f is continuous in $(0, 1]$. However, when we make x approach y to make $d_2(f(x), f(y))$ smaller than a constant, we observe that for each $\epsilon > 0$, the smaller y is, the smaller $d_1(x, y) = |x - y|$ should be. Thus, no such $\delta(\epsilon)$ for which $d_1(x, y) < \delta(\epsilon) \Rightarrow d_2(f(x), f(y)) < \epsilon$ exists if $\delta(\epsilon)$ does not depend on $x, y \in M$ (Fig. 2.1).

Proposition 8 *A continuous function over a bounded closed set is uniformly continuous.*

Fig. 2.1 The function $f(x) = 1/x$ is not uniformly continuous over $(0, 1]$. To make the $|f(x) - f(y)|$ value smaller than a constant, we need to make the $|x - y|$ value smaller when x, y are close to 0 (red lines) than that when x, y are far away from 0 (blue lines). Thus, $\delta > 0$ depends on the locations of x, y

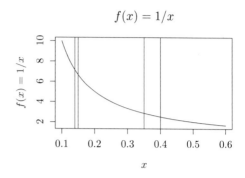

Proof: Let $f : E \to \mathbb{R}$ be a continuous function defined over a bounded closed set M. Because the function f is continuous, for an arbitrary $\epsilon > 0$, there exists a $\Delta(z)$ for each $z \in M$ such that

$$d_1(x, z) < \Delta(z) \Rightarrow d_2(f(x), f(z)) < \epsilon. \tag{2.2}$$

From Proposition 7, we can prepare a finite number of neighborhoods to cover M. Let $U_1, \ldots, U_m$ be neighborhoods with centers $z_1, \ldots, z_m$ and radii $\Delta(z_1)/2, \ldots, \Delta(z_m)/2$. Suppose that we choose $x, y \in M$ such that $d_1(x, y) < \delta := \frac{1}{2} \min_{1 \le i \le m} \Delta(z_i)$. Because x belongs to one of the $U_1, \ldots, U_m$, without loss of generality, we assume that $x \in U_i$. From the distance property, we have

$$d_1(x, z_i) < \frac{1}{2}\Delta(z_i) < \Delta(z_i)$$

$$d_1(y, z_i) \le d_1(x, y) + d_1(x, z_i) < \Delta(z_i).$$

Combining these inequalities, from the assumption that f is continuous and (2.2), we have That

$$d_2(f(x), f(y)) \le d_2(f(x), f(z_i)) + d_2(f(y), f(z_i)) < \epsilon + \epsilon = 2\epsilon.$$

Since $\epsilon > 0$ is arbitrary and δ does not depend on x, y, f is uniformly continuous. $\square$

Example 30 We can prove that "a definite integral exists for a continuous function defined over a closed interval $[a, b]$" by virtue of Proposition 8. If we divide $a < b$ into n equal-length segments as $a = x_0 < \ldots < x_n = b$, then for an arbitrary $\epsilon > 0$, we require

$$\left\{ \sum_{i=1}^{n} \frac{b-a}{n} \sup_{x_{i-1} < x < x_i} f(x) \right\} - \left\{ \sum_{i=1}^{n} \frac{b-a}{n} \inf_{x_{i-1} < x < x_i} f(x) \right\} < \epsilon$$

to define the definite integral. Because we assume that f is uniformly continuous, the condition $|f(x) - f(y)| < \epsilon/(b - a)$, $x, y \in [x_{i-1}, x_i]$ is satisfied if we make $\delta = x_i - x_{i-1} = \dfrac{b - a}{n}$ smaller (i.e., we make n larger).

2.2 Linear Spaces and Inner Product Spaces

We say that a set V is a linear space[3] if it satisfies the following conditions: for $x, y \in V$ and $\alpha \in \mathbb{R}$,

1. $x + y \in V$;
2. $\alpha x \in V$.

Example 31 If we define the sum of $x = [x_1, \ldots, x_d]$, $y = [y_1, \ldots, y_d] \in \mathbb{R}^d$ and the multiplication by a constant $\alpha \in \mathbb{R}$ as $x + y = [x_1 + y_1, \ldots, x_d + y_d]$ and $\alpha x = [\alpha x_1, \ldots, \alpha x_d]$, respectively, then the d-dimensional Euclidean space $\mathbb{R}^d$ forms a linear space.

Example 32 (L^2 Space) The set $L^2[0, 1]$ of functions $f : [0, 1] \to \mathbb{R}$ for which $\int_0^1 \{f(x)\}^2 dx$ takes a finite value is a linear space because

$$\int_0^1 \{f(x) + g(x)\}^2 dx \leq 2 \int_0^1 f(x)^2 dx + 2 \int_0^1 g(x)^2 dx < \infty$$

$$\int_0^1 \{\alpha f(x)\}^2 dx = \alpha^2 \int_0^1 f(x)^2 dx < \infty$$

for $\alpha \in \mathbb{R}$ when $\int_0^1 \{f(x)\}^2 dx < \infty$ and $\int_0^1 g(x)^2 dx < \infty$.

Let V be a linear space. We say that any bivariate function $\langle \cdot, \cdot \rangle : V \times V \to \mathbb{R}$ that obeys the four conditions below is an inner product:

1. $\langle x, x \rangle \geq 0$;
2. $\langle \alpha x + \beta y, z \rangle = \alpha \langle x, z \rangle + \beta \langle y, z \rangle$;
3. $\langle x, y \rangle = \langle y, x \rangle$;
4. $\langle x, x \rangle = 0 \iff x = 0$

for $x, y, z \in V$ and $\alpha, \beta \in \mathbb{R}$.

Example 33 For the linear space $\mathbb{R}^d$ in Example 31,

$$\langle x, y \rangle = \sum_{i=1}^{d} x_i y_i$$

---

[3] The same as a vector space.

for $x = [x_1, \ldots, x_d]$, and $y = [y_1, \ldots, y_d] \in \mathbb{R}^d$ is an inner product because the first three conditions are obvious, and the last condition holds because

$$\langle x, x \rangle = 0 \Longleftrightarrow \sum_{i=1}^{d} x_i^2 = 0 \Longleftrightarrow x = 0.$$

For a given linear space, we need to choose its inner product. We say that a linear space equipped with an inner product is an inner product space.

Example 34 (Inner Product of the L^2 Space) Let $L^2[0, 1]$ be the linear space in Example 32. The bivariate function

$$\langle f, g \rangle = \int_0^1 f(x)g(x)dx$$

$f, g \in L^2[0, 1]$ is not an inner product because the last condition fails. If $f(1/2) = 1$ and $f(x) = 0$ for $x \neq 1/2$, then we have

$$\langle f, f \rangle = \int_0^1 f(x)^2 dx = 0.$$

Strictly speaking, we construct such an inner product, identifying[4] $f, g \in L^2$ if and only if $\int_0^1 \{f(x) - g(x)\}^2 dx = 0$, which may rarely be noticed.

Let V be a linear space. We say that any function $\| \cdot \| : V \to \mathbb{R}$ that obeys the four conditions below is a norm.

1. $\|x\| \geq 0$;
2. $\|av\| = |a|\|x\|$;
3. $\|x + y\| \leq \|x\| + \|y\|$ (triangle inequality);
4. $\|x\| = 0 \Longleftrightarrow x = 0$

for $x, y \in V$ and $a \in \mathbb{R}$.

If we define the norm of V, then we can construct a metric space with the distance $d(x, y) = \|x - y\|$. If we define the inner product of V, the function

$$\|x\| = \langle x, x \rangle^{1/2} \tag{2.3}$$

satisfies the four conditions of a norm, and we call this norm a norm induced by an inner product.

In Examples 32 and 34, we introduced L^2 over $E = [0, 1]$ using the Riemann integral. However, in general, we define $L^2(E, \mathcal{F}, \mu)$ according to the set of $f : E \to \mathbb{R}$ for which

---

[4] We define the equivalent relation $\sim$ such that $f - g \in \{h| \int_0^1 \{h(x)\}^2 dx = 0\} \Longleftrightarrow f \sim g$ and construct the inner product for the quotient space $L^2/\sim$.

$$\int_E f^2 d\mu \qquad (2.4)$$

is finite[5] for the measure space $(E, \mathcal{F}, \mu)$.

Example 35 (Uniform Norm) The set of continuous functions over $[a, b]$ forms a linear space. The uniform norm defined by

$$\|f\| := \sup_{x\in[a,b]} |f(x)| , \quad f \in C[a, b]$$

is not induced by an inner product but satisfies the norm conditions:

$$\|f\| = 0 \iff f(x) = 0 , \quad x \in [a, b] .$$

In this book, we often use the Cauchy-Schwarz inequality:

$$|\langle x, y \rangle| \leq \|x\| \, \|y\| \qquad (2.5)$$

for $x, y \in V$. Apparently, (2.5) holds for $y = 0$. If $y \neq 0$, since the inner product of y and $z := x - \dfrac{\langle x, y \rangle}{\|y\|^2} y$ is zero, we have

$$\|x\|^2 = \|z + \frac{\langle x, y \rangle}{\|y\|^2} y\|^2 = \|z\|^2 + \|\frac{\langle x, y \rangle}{\|y\|^2} y\|^2 \geq \frac{\langle x, y \rangle^2}{\|y\|^2} ,$$

where the equality holds if and only if $z = 0$, which occurs exactly when one of x, y is a constant multiplication of the other. Moreover, we can examine the triangle norm inequality via (2.5):

$$\|x + y\|^2 = \|x\|^2 + 2|\langle x, y \rangle| + \|y\|^2 \leq \|x\|^2 + 2\|x\| \, \|y\| + \|y\|^2 = (\|x\| + \|y\|)^2 .$$

Because we did not use the last condition of inner products in deriving the Cauchy-Schwarz inequality, we may apply this inequality to any bivariate function that satisfies the first three conditions.

Using Cauchy-Schwarz's inequality, we can prove the continuity of an inner product.

Proposition 9 (Continuity of an Inner Product) *Let $\{x_n\}, \{y_n\}$ be sequences in a linear space V. For $x, y \in V$, if $x_n \to x$ and $y_n \to y$ as $n \to \infty$, then $\langle x_n, y_n \rangle \to \langle x, y \rangle$, where we denote by $\| \cdot \|$ the norm induced by the inner product of V.*

Proof: The proposition follows from $\|x_n\| \leq \|x\| + \|x_n - x\| \to \|x\| \, (n \to \infty)$ and

$$|\langle x_n, y_n \rangle - \langle x, y \rangle| \leq |\langle x_n, y_n - y \rangle| + |\langle x_n - x, y \rangle| \leq \|x_n\| \cdot \|y_n - y\| + \|x_n - x\| \cdot \|y\| \to 0.$$

$\square$

---

[5] We often abbreviate $(E, \mathcal{F}, \mu)$ or specify the interval as $[a, b]$ (as in $L^2[a, b]$).

2.3 Hilbert Spaces

We say that a vector space in which a norm is defined and the distance is complete is a Banach space. Hereafter, we denote by $C(E)$ the set of continuous functions defined over E.

Example 36 If $p \geq 1$, the vector space $\mathbb{R}^p$ is complete under the standard inner product and is a Hilbert space. On the other hand, the vector space $\mathbb{Q}^p$ is not complete under the standard inner product and does not make a Hilbert space in this case.

Example 37 The set of continuous functions $[0, 1] \rightarrow \mathbb{R}$ forms a linear space $C[0, 1]$. We consider the function

$$
f_n(t) := \begin{cases} 0, & 0 \leq t \leq \dfrac{1}{2} \\ n(t - \tfrac{1}{2}), & \dfrac{1}{2} < t < \dfrac{1}{2} + \dfrac{1}{n} \\ 1, & \dfrac{1}{2} + \dfrac{1}{n} \leq t \leq 1 . \end{cases}
$$

For $m \geq n$, we have

$$
\begin{aligned}
\|f_n - f_m\|_2^2 &= \int_0^1 |f_n(t) - f_m(t)|^2 dt = \int_{\frac{1}{2}}^{\frac{1}{2}+\frac{1}{n}} |f_n(t) - f_m(t)|^2 dt \\
&= \int_{\frac{1}{2}}^{\frac{1}{2}+\frac{1}{m}} [n(t - \tfrac{1}{2}) - m(t - \tfrac{1}{2})]^2 dt + \int_{\frac{1}{2}+\frac{1}{m}}^{\frac{1}{2}+\frac{1}{n}} [n(t - \tfrac{1}{2}) - 1]^2 dt \\
&= \frac{(n-m)^2}{3m^3} - \frac{(n-m)^3}{3m^3 n} = \frac{(n-m)^2}{3m^2 n} < \frac{1}{3n} \rightarrow 0 .
\end{aligned}
$$

Thus, $\{f_n\}$ is a Cauchy sequence in $C[0, 1]$ (Fig. 2.2). However, f_n converges to a function that is not continuous:

$$
f(t) := \begin{cases} 0, \ 0 \leq t \leq \tfrac{1}{2} \\ 1, \ \tfrac{1}{2} < t \leq 1 \end{cases}
$$

$$
\|f_n - f\|^2 = \int_0^1 \|f_n(t) - f(t)\|^2 dt = \int_{\frac{1}{2}}^{\frac{1}{2}+\frac{1}{n}} [n(t - \tfrac{1}{2}) - 1]^2 dt = \frac{1}{3n} \rightarrow 0 .
$$

As we have seen, $C[a, b]$ is not complete w.r.t. the L^2 norm. However, it is complete w.r.t. the uniform norm:

Proposition 10 $C[a, b]$ *is complete w.r.t. the uniform norm.*

Proof: Let $\{f_n\}$ be a Cauchy sequence in $C[a, b]$, which means that

$$
\sup_{m,n \geq N} \sup_{x \in [a,b]} |f_m(x) - f_n(x)| \rightarrow 0 \tag{2.6}
$$

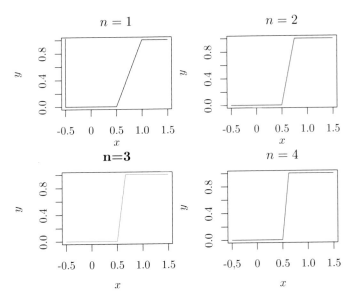

Fig. 2.2 We illustrate Example 37 when $f_n \to f$. The function is continuous for finite values of n but is not continuous as $n \to \infty$

as $N \to \infty$. Then, the real sequence $\{f_n(x)\}$ is a Cauchy sequence for each $x \in [a, b]$ and converges to a real value (Proposition 6). If we define the function $f(x)$ with $x \in E$ by $\lim_{n \to \infty} f_n(x)$, $\sup_{n \geq N} f_n(x)$ and $\inf_{n \geq N} f_n(x)$ converge to $f(x)$ from above and from below, respectively. From (2.6), we see that

$$|f_N(x) - f(x)| \leq \sup_{n \geq N} f_n(x) - \inf_{n \geq N} f_n(x) = \sup_{m,n \geq N} |f_m(x) - f_n(x)|$$

uniformly converges to 0 for an arbitrary $x \in [a, b]$, which implies that $C[a, b]$ is complete. □

Because any inner product does not induce the uniform norm, $C[a, b]$ is a Banach space but is not a Hilbert space.

Proposition 11 $C[a, b]$ *is separable w.r.t. the uniform norm.*

For the proof, we use the Stone-Weierstrass theorem (Proposition 12) [30, 31, 34, 35]. The term algebra is used to denote the linear space A that defines associative "·" and commutative "+" properties and satisfies

$$x \cdot (y + z) = x \cdot y + x \cdot z,$$

$$(y + z) \cdot x = y \cdot x + z \cdot x,$$

$$\alpha(x \cdot y) = (\alpha x) \cdot y$$

for x, y, $z \in A$ and $\alpha \in \mathbb{R}$, where the first two properties are identical if $\cdot$ is commutative. The general theory may be complex, but we only suppose that $+$ and $\cdot$ are the standard addition and multiplication operations and that A is either a polynomial or continuous function.

Example 38 (Polynomial Ring) Let $+,\cdot$ be the standard commutative addition and multiplication operations. The polynomial ring $\mathbb{R}[x, y, z]$ is a set of polynomials with indeterminates (variables) x, y, z and is an algebra with $\mathbb{R}$ commutative coefficients. $\mathbb{R}[x, y, z]$ is a linear space, and if two elements belong to $\mathbb{R}[x, y, z]$, multiplication by $\mathbb{R}$ and addition among the elements belong to $\mathbb{R}[x, y, z]$. Moreover, the three laws follow for the elements in $\mathbb{R}[x, y, z]$.

Proposition 12 (Stone-Weierstrass [30, 31, 34, 35]) *Let E and A be a compact set and an algebra, respectively. Under the following conditions, A is dense in C(E).*

1. *No $x \in E$ exists such that $f(x) = 0$ for all $f \in A$.*
2. *For each pair x, $y \in E$ with $x \neq y$, $f \in A$ exists such that $f(x) \neq f(y)$.*

We refer to Proposition 12 several times. Although we abbreviate its proof in this book, please follow it because it contains no complicated derivation processes. We can use this proposition to prove that a neural network can approximate any continuous function.

Proof of Proposition 11: A set A of polynomials with indeterminate x and real coefficients satisfies the two conditions in Proposition 12. Hence, A is dense in $C[a, b]$. Furthermore, if we restrict the coefficients of A to $\mathbb{Q}$, then A is a countable set, which means that $C[a, b]$ is separable. $\qquad\square$

Proposition 13 $C[a, b]$, $a < b$, is dense in $L^2[a, b]$.

For the proof, see the Appendix at the end of this chapter. We say that a function in the form

$$\sum_{k=1}^{m} h(B_k) I(B_k) \tag{2.7}$$

with exclusive $B_1, \ldots, B_m \in \mathcal{F}$ and $h : \mathcal{F} \to \mathbb{R}_{\geq 0}$ is a simple function. Equation 1.7 in Chap. 1 approximates this function by using a simple function. The proof in the Appendix shows that a simple function approximates an arbitrary $f \in L^2$ and that a continuous function approximates an arbitrary simple function. $\qquad\square$

Proposition 14 (Riesz-Fischer) L^2 *is complete*[6]. *In other words, L^2 is a Banach space and a Hilbert space.*

---

[6] L^p is complete for any $p \geq 1$, although we abbreviate this proof.

The outline of the proof is as follows; see the Appendix for details. It is sufficient to show that "$\{f_n\}$ is a Cauchy sequence in $L^2 \Rightarrow$ there exists an $f \in L^2$ such that $\|f_n - f\| \to 0$". We define the f to which the Cauchy sequence $\{f_n\}$ in L^2 converges and derive $\|f_n - f\| \to 0$ and $f \in L^2$.

1. Let $\{f_n\}$ be an arbitrary Cauchy sequence in L^2.
2. There exists $\{n_k\}$ such that $\|\sum_{k=1}^{\infty} |f_{n_{k+1}} - f_{n_k}|\|_2 < \infty$.
3. We show the existence of $f : E \to \mathbb{R}$ such that $\mu\{x \in E \mid \lim_{k \to \infty} f_{n_k}(x) = f(x)\} = \mu(E)$.
4. We show that $\|f_n - f\| \to 0$ and $f \in L^2[a, b]$.

□

Let V and $\langle \cdot, \cdot \rangle$ be an inner product space and its inner product, respectively. We say that $x, y \in V$ are orthogonal if $\langle x, y \rangle = 0$ and that a sequence $\{e_j\}$ in V is orthonormal if $\|e_j\| = 1$ for each j and each pair e_i, e_j $(i \neq j)$ is orthogonal. Moreover, we say that $\{e_j\}$ is an orthonormal basis of V if $\{e_j\}$ is an orthonormal sequence and each $x \in V$ can be expressed by $x = \sum_{j=1}^{\infty} \alpha_j e_j$ using the real α_j.

Proposition 15 *For an orthonormal sequence $\{e_j\}$ in a Hilbert space H, we have the following properties :*

1. $\displaystyle\sum_{i=1}^{\infty} \langle x, e_i \rangle^2 \leq \|x\|^2$ *(Bessel's inequality).*

2. $\displaystyle\sum_{i=1}^{\infty} \langle x, e_i \rangle e_i$ *converges.*

3. $\displaystyle\sum_{i=1}^{\infty} \alpha_i e_i$ *converges* $\Longleftrightarrow \sum_{i=1}^{\infty} \alpha_i^2 < \infty$.

4. $\displaystyle y = \sum_{i=1}^{\infty} \alpha_i e_i \Rightarrow \alpha_i = \langle y, e_i \rangle$.

Proof: See the Appendix at the end of this chapter.

Suppose that A is a subset of a linear space V equipped with a norm, and let $\overline{\text{span}(A)}$ be the closure of span(A), which is a linear combination of the elements in A. Suppose that $\{x_n\}$ is a sequence in a Hilbert space and that each x_n is orthogonal. Then, the sequence $\{e_n\}$ constructed below is orthonormal

$$v_i := x_i - \sum_{j=1}^{i-1} \langle x_i, e_j \rangle e_j , \quad e_i = v_i / \|v_i\| , \quad i = 1, 2, \ldots$$

and satisfies $\overline{\text{span}\{x_n\}} = \overline{\text{span}\{e_n\}}$ (Gram-Schmidt).

Proposition 16 *Let* $\{e_j\}$ *be an orthonormal sequence in a Hilbert space H. The following conditions are equivalent:*

1. $\{e_i\}$ *is an orthonormal sequence in H.*
2. *For an arbitrary* $x \in H$, $\langle x, e_i \rangle = 0$, $k = 1, 2, \ldots, \Rightarrow x = 0$.
3. span$\{e_i\}$ *is dense in H.*
4. *The equality (Parseval) of Bessel's inequality (the first of Proposition 15).*
5. *For arbitrary* $x, y \in H$, $\langle x, y \rangle = \sum_{k=1}^{\infty} \langle x, e_k \rangle \langle y, e_k \rangle$.
6. *For an arbitrary* $x \in H$, $x = \sum_{j=1}^{\infty} \langle x, e_j \rangle e_j$.

Proof: See the Appendix at the end of this chapter.

Example 39 (Fourier Series Expansion)By approximating $f \in L^2[-\pi, \pi]$ by

$$f_m(x) = a_0 + \sum_{n=1}^{m} (a_n \cos nx + b_n \sin nx) \tag{2.8}$$

and computing the $\{a_n\}$, $\{b_n\}$ that minimizes $\|f - f_m\|$, we obtain $\|f - f_m\| \to 0$ as $m \to \infty$. However, $\|f - f_m\| = 0$ does not occur for any m. We see that $f \notin \text{span}(A)$ for $A := \{1, \cos x, \sin x, \cos 2x, \sin 2x, \cdots\}$ and that span(A) is not a closed set. We add all the $f \notin \text{span}(A)$ such that $\|f - f_m\| \to 0$ (convergence points) for span(A) to obtain the closure $\overline{\text{span}(A)}$. Moreover, span$(A)$ is dense in $L^2[-\pi, \pi]$. Then,

$$\{\frac{1}{\sqrt{2\pi}}, \frac{\cos x}{\sqrt{\pi}}, \frac{\sin x}{\sqrt{\pi}}, \frac{\cos 2x}{\sqrt{\pi}}, \frac{\sin 2x}{\sqrt{\pi}}, \cdots\}$$

which are the $\{1, \cos x, \sin x, \cos 2x, \sin 2x, \cdots\}$ divided by their norms, forming an orthonormal basis of the Hilbert space $L^2[-\pi, \pi]$ that consists of the functions expressed by the Fourier series as in (2.8), where we regard $\langle f, g \rangle :=$ $\int_{-\pi}^{\pi} f(x)g(x)dx$ as the inner product of $f, g \in H$, and we use $\int_{-\pi}^{\pi} \cos mx \sin nx dx = 0$ and $\int_{-\pi}^{\pi} \cos^2 mx dx = \int_{-\pi}^{\pi} \sin^2 nx dx = \pi$ for $m, n > 0$.

Proposition 17 *For a Hilbert space H, being separable is equivalent to having an orthonormal basis.*

Proof: If $\{e_j\}$ forms an orthonormal basis of H, we can express an arbitrary $x \in H$ as $x = \sum_{j=1}^{\infty} \langle x, e_j \rangle e_j$. $E := \{x \in H : \langle x, e_j \rangle \in \mathbb{Q}, j = 1, 2, \ldots\}$ is dense and countable. In fact, $\{\langle x, e_j \rangle\}$ and $\{e_j\}$ are countable sets, as is E, that is, the combination of these two sets, which means that H is separable. On the other hand, if H is separable, by extracting linearly independent elements from H that are dense and countable via Gram-Schmidt's method, we can construct an orthonormal basis $\{e_n\}$. Thus, we see that any linear combination of these elements is dense in H. From the third part of Proposition 16, $\{e_i\}$ is an orthonormal basis of H. $\square$

Proposition 17 implies the following:

Proposition 18 $L^2[a, b]$ *is separable under the* L^2 *norm.*

In this book, we assume that the Hilbert space we deal with is separable.

2.4 Projection Theorem

Let V and M be a linear space equipped with an inner product and its subspace, respectively. We define the orthogonal complement of M as

$$M^\perp := \{x \in V : \langle x, y \rangle = 0 \text{ for all } y \in M\} .$$

For subspaces M_1 and M_2 of V, we write the direct sum of M_1, M_2 as $M_1 + M_2$. In particular, when M_1, M_2 are orthogonal to each other, i.e.,

$$x_1 \perp x_2 , \ x_1 \in M_1 , \ x_2 \in M_2 , \tag{2.9}$$

we write it as $M_1 \oplus M_2 := \{x_1 + x_2 : x_1 \in M_1, x_2 \in M_2\}$.

Proposition 19 (Projection Theorem) *Let M be a closed subset of a Hilbert space H. Then, for an arbitrary $x \in H$, there exists a $y \in M$ that minimizes $\|x - y\|$. Moreover such a y satisfies*

$$\langle x - y, z \rangle = 0 , \ z \in M \tag{2.10}$$

and is unique.

Proof: Given an $x \in H$, we consider a $y \in M$ such that $x = y + (x - y)$, $y \in M$, and $x - y \in M^\perp$. For the proof, we utilize the following steps:

1. Show that a sequence $\{y_n\}$ in M such that

$$\lim_{n \to \infty} \|x - y_n\|^2 = \inf_{y \in M} \|x - y\|^2$$

   is a Cauchy sequence.
2. Demonstrate the existence of a $y \in M$ such that $y_n \to y \in M$.
3. Show that $2a \langle x - y, z - y \rangle \leq a^2 \|z - y\|^2$ for arbitrary $0 < a < 1$ and $z \in M$, and derive a contradiction in the inequality when we assume that $\langle x - y, z - y \rangle > 0$.
4. Show that $\langle x - y, z \rangle \leq 0$.
5. Substitute $-z$ instead of z to obtain the proposition.

For details, see the Appendix at the end of this chapter.   □

Equation (2.10) implies that an arbitrary $x \in H$ can be uniquely decomposed into $x = y + (x - y)$ with $y \in M$ and $x - y \in M^\perp$, which means that

$$H = M \oplus M^\perp . \tag{2.11}$$

Example 40 For a positive definite kernel $k : E \times E \to \mathbb{R}$ and each $x \in E$, we regard $k(x, \cdot) : E \to \mathbb{R}$ as a function over E. In general, the space spanned by $\{k(x, \cdot)\}_{x \in E}$ and its closure $H := \overline{\text{span}(\{k(x, \cdot)\}_{x \in E})}$ is a linear space. We show that the inner product is $\langle k(x, \cdot), k(y, \cdot) \rangle_H = k(x, y)$ and demonstrate its completeness in Chap. 3. For $x_1, \ldots, x_N \in E$, $M := \mathrm{span}(\{k(x_i, \cdot)\}_{i=1}^N)$ forms a finite-dimensional linear space, and H can be written as (2.11) by using

$$M^\perp = \{f \in H \,|\, \langle f, k(x_i, \cdot) \rangle_H = 0,\ i = 1, \ldots, N\}.$$

If $f = f_1 + f_2$, for $f_1 \in M$ and $f_2 \in M^\perp$, we have

$$\|f\|_H^2 = \|f_1\|_H^2 + \|f_2\|_H^2 + 2\langle f_1, f_2 \rangle_H = \|f_1\|_H^2 + \|f_2\|_H^2 \geq \|f_1\|_H^2,$$

which is used in Chap. 4 for kernel calculations.

Proposition 20 Let H and M be a Hilbert space and its subset,[7] respectively. Then, we have the following:

1. $M^\perp$ is a closed subset of H.
2. $M \subseteq (M^\perp)^\perp$.
3. If M is a subspace, then $(M^\perp)^\perp = \overline{M}$, where $\overline{M}$ is a closure of the set M.

Proof: For the first item, we see that $M^\perp$ is a subspace. From continuity of an inner product (Proposition 9), if $x_n \to x$ as $n \to \infty$ for a sequence $\{x_n\}$ in $M^\perp$, then we have for $a \in M$,

$$\langle x, a \rangle = \lim_{n \to \infty} \langle x_n, a \rangle = 0,$$

which means that $M^\perp$ is closed. The second item is due to

$$x \in M \Longrightarrow \langle x, y \rangle = 0,\ y \in M^\perp \Longrightarrow x \in (M^\perp)^\perp.$$

For the third item, from the first two properties, taking the closure on both sides of $M \subseteq (M^\perp)^\perp$ yields $\overline{M} \subseteq (M^\perp)^\perp$. From Proposition 19, an arbitrary $x \in (M^\perp)^\perp$ can be written as $y \in \overline{M} \cap (M^\perp)^\perp = \overline{M}$ and $z \in \overline{M}^\perp \cap (M^\perp)^\perp$. However, we have $\overline{M}^\perp \cap (M^\perp)^\perp \subseteq M^\perp \cap (M^\perp)^\perp = \{0\}$, which means that $z = 0$, and we obtain the third item.   □

2.5 Linear Operators

Let X_1, X_2 be linear spaces with norms of $\|\cdot\|_1$, $\|\cdot\|_2$, respectively, and let $T : X_1 \to X_2$ be the map that linearly transforms an element in X_1 to an element in X_2. We call such a T a linear operator. We define an image and a kernel by

$$\mathrm{Im}(T) := \{Tx : x \in X_1\} \subseteq X_2$$

---

[7] This is not necessarily a subspace.

and

$$\text{Ker}(T) := \{x \in X_1 : Tx = 0\} \subseteq X_1 ,$$

and we call the dimensionality of $\text{Im}(T)$ the rank of T. We say that the linear operator $T : X_1 \to X_2$ is bounded if for each $x \in X_1$, there exists a constant $C > 0$ such that

$$\|Tx\|_2 \le C\|x\|_1 .$$

We write the set of such T's as $B(X_1, X_2)$. In particular, we write $B(X_1, X_2)$ as $B(X)$ when $X_1 = X_2 = X$.

Proposition 21 *A linear operator T is bounded if and only if T is uniformly continuous.*

Proof: If T is uniformly continuous, then there exists a $\delta > 0$ such that $\|x\|_1 \le \delta \Rightarrow \|Tx\|_2 \le 1$. Since $\|\dfrac{\delta x}{\|x\|}\| \le \delta$, we have

$$\|Tx\|_2 = \|T(\frac{\delta x}{\|x\|_1})\|_2 \frac{\|x\|_1}{\delta} \le \frac{\|x\|_1}{\delta}$$

for any $x \ne 0$. On the other hand, if T is bounded, there exists a constant C that does not depend on $x \in X_1$ such that

$$\|T(x_n - x)\|_2 \le C\|x_n - x\|_1$$

for any $\{x_n\}$ and an $x \in X_1$ such that $x_n \to x$ as $n \to \infty$. □

Hereafter, we define the operator norm of $T \in B(X_1, X_2)$ by

$$\|T\| := \sup_{x \in X_1, \|x\|_1 = 1} \|Tx\|_2 . \qquad (2.12)$$

Thus, for an arbitrary $x \in X_1$, we have

$$\|Tx\|_2 \le \|T\| \|x\|_1 .$$

Example 41 Let $X_1 := \mathbb{R}^p$ and $X_2 := \mathbb{R}^q$. If the norm is the Euclidean norm, then we can write the linear operator $T : \mathbb{R}^p \to \mathbb{R}^q$ as $T : x \mapsto Bx$ by using some $B \in \mathbb{R}^{q \times p}$. If the matrix B is square, the norm $\|T\|$ is the square root of the maximum eigenvalue of the nonnegative definite matrix $A := B^\top B$.

$$\|T\|^2 = \max_{\|x\|=1} x^\top B^\top Bx = \max_{\|x\|=1} \|Bx\|^2 .$$

Example 42 For $K : [0, 1]^2 \to \mathbb{R}$, let

$$\int_0^1 \int_0^1 K^2(x, y)dxdy$$

be finite. We define the integral operator by the linear operator T in $L^2[0, 1]$ such that

$$(Tf)(\cdot) = \int_0^1 K(\cdot, x)f(x)dx \tag{2.13}$$

for $f \in L^2[0, 1]$. Note that (2.13) belongs to $L^2[0, 1]$ and that T is bounded: From

$$|(Tf)(x)|^2 \le \int_0^1 K^2(x, y)dy \int_0^1 f^2(y)dy = \|f\|_2^2 \int_0^1 K^2(x, y)dy ,$$

we have

$$\|Tf\|_2^2 = \int_0^1 |(Tf)(x)|^2 dx \le \|f\|_2^2 \int_0^1 \int_0^1 K^2(x, y)dxdy .$$

We call such a K an integral operator kernel and distinguish between the positive definite kernels we deal with in this book.

In particular, we call any linear operator with $X_2 = \mathbb{R}$ a linear functional.

Proposition 22 (Riesz's Representation Theorem)*Let H be a Hilbert space with an inner product $\langle \cdot, \cdot \rangle$ and a norm of $\| \cdot \|$, and let $T \in B(H, \mathbb{R})$. Then, there exists a unique $e_T \in H$ such that*

$$Tf = \langle f, e_T \rangle , \quad f \in H \tag{2.14}$$

and $\|T\| = \|e_T\|$.

Proof: See the Appendix at the end of this chapter.

Example 43 (RKHS) Let $x \in E$, and let $T_x : H \to \mathbb{R}$ be the map from $f \in H$ to $f(x)$. Then, T_x is linear because

$$T_x(af + bg) = (af + bg)(x) = af(x) + bg(x) = aT_x(f) + bT_x(g) .$$

We assume that T_x is bounded for each $x \in E$. Then, from Proposition 22, there exists a $k_x \in H$ such that

$$f(x) = T_x(f) = \langle f, k_x \rangle$$

$x \in E$, and $\|T_x\| = \|k_x\|$.

Proposition 23 (Adjoint Operator) *Let H_i be a Hilbert space with an inner product $\langle \cdot, \cdot \rangle_i$ for $i = 1, 2$ and $T \in B(H_1, H_2)$. Then, there exists a $T^* \in B(H_2, H_1)$ such that*

$$\langle Tx_1, x_2 \rangle_2 = \langle x_1, T^*x_2 \rangle_1 , \ x_1 \in H_1, \ x_2 \in H_2 .$$

Proof: If we fix $x_2 \in H_2$ and regard $\langle Tx_1, x_2 \rangle_2$ as a function of $x_1 \in H_1$, then from $x_1 \mapsto \langle Tx_1, x_2 \rangle_2 \leq \|x_1\|_2 \|x_2\|_2$, T is a bounded operator w.r.t. $x_1 \in H_1$. From Proposition 22, for each $x_2 \in H_2$, there exists $y_2(x_2) \in H_1$ such that $\langle Tx_1, x_2 \rangle_2 = \langle x_1, y(x_2) \rangle_1$. If we define $T^*x_2 = y_2(x_2)$, then T^* is a bounded linear map. The boundness property is due to

$$\|T^*x_2\|_1^2 = |\langle x_2, TT^*x_2 \rangle|_2 \leq \|T\| \ \|T^*x_2\|_1 \|x_2\|_2 .$$

□

We call the T^* in Proposition 23 the adjoint operator of T. In particular, if $T^* = T$, we call such an operator T self-adjoint.

Example 44 Let $H = \mathbb{R}^p$. We can express any $T \in B(H)$ by a square matrix $T \in \mathbb{R}^{p \times p}$. From

$$\langle Tx, y \rangle = x^\top T^\top y = \langle x, T^\top y \rangle ,$$

we see that the adjoint T^* is the transpose matrix of $T^\top$ and that T can be written as a symmetric matrix if and only if T is self-adjoint.

Example 45 For the integral operator of $L^2[0, 1]$ in Example 42, from Fubini's theorem, we have that

$$\langle Tf, g \rangle = \int_0^1 \int_0^1 K(x, y) f(x) g(y) dx dy = \langle f, \int_0^1 K(y, \cdot) g(y) dy \rangle ,$$

and $y \mapsto (T^*g)(y) = \int_0^1 K(x, y) g(x) dx$ is an adjoint operator. If the integral operator kernel K is symmetric, the operator T is self-adjoint.

2.6 Compact Operators

Let (M, d) and E be a metric space and a subset of M, respectively. If any infinite sequence in E contains a subsequence that converges to an element in E, then we say that E is sequentially compact. If $\{x_n\}$ has a subsequence that converges to x, then x is a convergence point of $\{x_n\}$.

Example 46 Let $E := \mathbb{R}$ and $d(x, y) := |x - y|$ for $x, y \in \mathbb{R}$. Then, E is not sequentially compact. In fact, the sequence $x_n = n$ has no convergence points. For $E = (0, 1]$, the sequence $x_n = 1/n$ converges to $0 \notin (0, 1]$ as $n \to \infty$, and the convergence point of any subsequence is only 0. Therefore, $E = (0, 1]$ is not sequentially compact.

Proposition 24 *Let (M, d) and E be a metric space and a subset of M, respectively. Then, E is sequentially compact if and only if E is compact.*

Proof: Many books on geometry deal with the proof of equivalence. See such books for the details of this proof.

In this section, we explain compactness by using the terminology of sequential compactness.

Let X_1, X_2 be linear spaces equipped with norms, and let $T \in B(X_1, X_2)$. We say that T is compact if $\{Tx_n\}$ contains a convergence subsequence for any bounded sequence $\{x_n\}$ in X_1.

Example 47 The orthonormal basis $\{e_j\}$ in a Hilbert space H is bounded because $\|e_j\| = 1$. However, for an identity map, we have that $\|e_i - e_j\| = \sqrt{2}$ for any $i \neq j$. Thus, the sequence $e_1, e_2, \ldots$ does not have any convergence points in H. Hence, the identity operator for any infinite-dimensional Hilbert space is not compact.

Proposition 25 *For any bounded linear operator T, the following hold:*

1. *If the rank is finite, then the operator T is compact.*
2. *If a sequence of finite-rank operators $\{T_n\}$ exists such that $\|T_n - T\| \to 0$ as $n \to \infty$, then T is compact[8].*

Proof: See the appendix at the end of this chapter.

Let H and $T \in B(H)$ be a Hilbert space and its bounded linear operator, respectively. If $\lambda \in \mathbb{R}$ and $0 \neq e \in H$ exist such that

$$Te = \lambda e, \tag{2.15}$$

then we say that λ and e are an eigenvalue and an eigenvector of T, respectively.

Proposition 26 *Let $T \in B(H)$ and $e_j \in \mathrm{Ker}(T - \lambda_j I)$ for $j = 1, 2, \ldots$. If the eigenvalues $\lambda_j \neq 0$ have different values, then*

1. *e_j is linearly independent.*
2. *If T is self-adjoint, then $\{e_j\}$ are orthogonal.*

Proof: See the appendix at the end of this chapter.

Example 48 Let $T \in B(H)$ be a compact operator. For each eigenvalue $\lambda \neq 0$, the eigenspace $\mathrm{Ker}(T - \lambda I)$ has a finite dimensionality. In fact, if $\mathrm{Ker}(T - \lambda I)$ is of infinite dimensionality for an eigenvalue $\lambda \neq 0$, then λ contains infinitely many eigenvectors e_j, and if we apply the operator T to them, then as in Example 47, $\{\lambda e_j\}$ does not have any convergence subsequence. Thus, T is not compact, which is a contradiction.

---

[8] It is known that the converse is true.

Example 49 For any $C > 0$, the absolute values of a finite number of eigenvalues λ_i for a compact operator T exceed C. Suppose that the absolute values of an infinite number of eigenvalues $\lambda_1, \lambda_2, \ldots$ exceed C. Let $M_0 := \{0\}$, $M_i := \text{span}\{e_1, \ldots, e_i\}$, $e_j \in \text{Ker}(T - \lambda_j I)$, $j = 1, 2, \ldots$, $i = 1, 2, \ldots$. Since the $\{e_1, \ldots, e_i\}$ are linearly independent, each $M_i \cap M_{i-1}^\perp$ is one dimensional for $i = 1, 2, \ldots$. Thus, if we define the orthonormal sequence $x_i \in \text{Ker}(T - \lambda_i I) \cap M_{i-1}^\perp, i = 1, 2, \ldots$ via Gram-Schmidt, then we have

$$\| T x_i - T x_k \|^2 = \| T x_i \|^2 + \| T x_k \|^2 \geq 2C^2$$

for $i > k$. Thus, $\{T x_i\}$ has no convergence subsequence.

Example 49 implies that the set of nonzero eigenvalues of T is countable. We summarize the above discussion and its implications below.

Proposition 27 *Let T be a self-adjoint compact operator of a Hilbert space H. Then, the set of nonzero eigenvalues of T is finite, or the sequence of eigenvalues converges to zero. Each eigenvalue has a finite multitude, and any pair of eigenvectors corresponding to different eigenvalues is an orthogonal pair. Let $\lambda_1, \lambda_2, \ldots$ be a sequence of eigenvalues such that $|\lambda_1| \geq |\lambda_2| \geq \cdots$, and let $e_1, e_2, \ldots$ be any corresponding eigenvectors that are orthogonal (orthogonalized eigenvectors via Gram-Schmidt) if they possess the same eigenvalue. Then, $\{e_j\}$ is the orthonormal basis of $\overline{\text{Im}(T)}$, and we can express T by*

$$T x = \sum_{j=1}^{\infty} \lambda_j \langle x, e_j \rangle e_j \tag{2.16}$$

for each $x \in H$.

Proof: We utilize the following steps, where the second item is equivalent to $(\text{Ker}(T))^\perp = \overline{\text{Im}(T)}$ because $T = T^*$.

1. Show that $H = \text{Ker}(T) \oplus (\text{Ker}(T))^\perp$.
2. Show that $(\text{Ker}(T))^\perp = \overline{\text{Im}(T^*)}$.
3. Show that $\overline{\text{span}\{e_j | j \geq 1\}} \subseteq \overline{\text{Im}(T)}$.
4. Show that $\overline{\text{span}\{e_j | j \geq 1\}} \supseteq \overline{\text{Im}(T)}$.

See the appendix at the end of this chapter.   □

We say that an operator T is nonnegative definite if

$$\langle T x, x \rangle = \langle \sum_{i=1}^{\infty} \lambda_i \langle x, e_i \rangle e_i, \sum_{j=1}^{\infty} \langle x, e_j \rangle e_j \rangle = \sum_{i=1}^{\infty} \lambda_i \langle x, e_i \rangle^2 \geq 0$$

for arbitrary $H \ni x = \sum_{i=1}^{\infty} \langle x, e_i \rangle e_i$; this condition is equivalent to $\lambda_1 \geq 0, \lambda_2 \geq 0, \ldots$.

Proposition 28 *If T is nonnegative definite, we have*

$$\lambda_k = \max_{e \in \text{span}\{e_1, \dots, e_{k-1}\}^{\perp}} \frac{\langle Te, e \rangle}{\|e\|^2} \tag{2.17}$$

which expresses the maximum value over the Hilbert space H when $k = 1$.

Proof: The claim follows from (2.16) and $\lambda_j \geq 0$:

$$\max_{e \in \{e_1, \dots, e_{k-1}\}^{\perp} \|e\|=1} \langle Te, e \rangle = \max_{\|e\|=1} \sum_{j=k}^{\infty} \lambda_j \langle e, e_j \rangle^2 = \lambda_k .$$

$\square$

Let H_1, H_2 be Hilbert spaces, $\{e_i\}$ an orthonormal basis of H_1, and $T \in B(H_1, H_2)$. If

$$\sum_{i=1}^{\infty} \|Te_i\|^2$$

takes a finite value, we say that T is a Hilbert-Schmidt (HS) operator, and we write the set of HS operators in $B(H_1, H_2)$ as $B_{HS}(H_1, H_2)$.

We define the inner product of $T_1, T_2 \in B_{HS}(H_1, H_2)$ and the HS norm of $T \in B_{HS}(H_1, H_2)$ by $\langle T_1, T_2 \rangle_{HS} := \sum_{j=1}^{\infty} \langle T_1 e_j, T_2 e_j \rangle_2$ and

$$\|T\|_{HS} := \langle T, T \rangle_{HS}^{1/2} = \left\{ \sum_{i=1}^{\infty} \|Te_i\|_2^2 \right\}^{1/2},$$

respectively.

Proposition 29 *The HS norm value of $T \in B(H_1, H_2)$ does not depend on the choice of orthonormal basis $\{e_i\}$.*

Proof: Let $\{e_{1,i}\}$, $\{e_{2,j}\}$ be arbitrary orthonormal bases of Hilbert spaces H_1, H_2, and let $T_1, T_2 \in B(H_1, H_2)$. Then, for $T_k e_{1,i} = \sum_{j=1}^{\infty} \langle T_k e_{1,i}, e_{2,j} \rangle_2 e_{2,j}$, $T_k^* e_{2,j} = \sum_{i=1}^{\infty} \langle T_k^* e_{2,j}, e_{1,i} \rangle_1 e_{1,i}$, and $k = 1, 2$, we have

$$\sum_{i=1}^{\infty} \langle T_1 e_{1,i}, T_2 e_{1,i} \rangle_2 = \sum_{i=1}^{\infty} \sum_{j=1}^{\infty} \langle T_1 e_{1,i}, e_{2,j} \rangle_2 \langle T_2 e_{1,i}, e_{2,j} \rangle_2$$

$$= \sum_{i=1}^{\infty} \sum_{j=1}^{\infty} \langle e_{1,i}, T_1^* e_{2,j} \rangle_1 \langle e_{1,i}, T_2^* e_{2,j} \rangle_1 = \sum_{i=1}^{\infty} \langle T_1^* e_{2,j}, T_2^* e_{2,j} \rangle_1 ,$$

which means that both sides do not depend on the choices of $\{e_{1,i}\}$, $\{e_{2,j}\}$. In particular, if $T_1 = T_2 = T$, we see that $\|T\|_{HS}^2$ does not depend on the choices of $\{e_{1,i}\}$, $\{e_{2,j}\}$.

$\square$

Proposition 30 *An HS operator is compact.*

Proof: Let $T \in B(H_1, H_2)$ be an HS operator, $x \in H_1$, and

$$T_n x := \sum_{i=1}^{n} \langle Tx, e_{2i} \rangle_2 e_{2i} \, ,$$

where $\{e_{2i}\}$ is an orthonormal basis of H_2. Since the image of T_n is of finite dimensionality, T_n is compact. Thus, from the second item of Proposition 25, it is sufficient to show that $\|T - T_n\| \to 0$ as $n \to \infty$. However, since $(T - T_n)x = \sum_{i=n+1}^{\infty} \langle Tx, e_{2,i} \rangle_2 e_{2,i}$, we have that when $\|x\|_1 \leq 1$,

$$\|(T - T_n)x\|_2^2 = \sum_{i=n+1}^{\infty} \langle Tx, e_{2,i} \rangle_2^2 = \sum_{i=n+1}^{\infty} \langle x, T^* e_{2,i} \rangle_1^2 \leq \sum_{i=n+1}^{\infty} \|T^* e_{2,i}\|^2 \, .$$

Because T^* is an HS operator, the right-hand side converges to zero, where T is an HS operator if and only if T^* is an HS operator due to the derivation in Proposition 29. $\qquad\square$

Example 50 When an operator is expressed by a matrix $T = (T_{i,j})$ such that $T \in B(\mathbb{R}^m, \mathbb{R}^n)$, $m, n \geq 1$, the HS norm becomes the squared sum of the mn elements of this matrix. In fact, if T is expressed by a matrix $\mathbb{R}^{n \times m}$, then the HS norm is the Frobenius norm:

$$\|T\|_{HS}^2 = \sum_{i=1}^{n} \|T e_{X,i}\|^2 = \sum_{j=1}^{n} \|T^* e_{Y,j}\|^2 = \sum_{i=1}^{m} \sum_{j=1}^{n} T_{i,j}^2 \, ,$$

where $e_{X,i} \in \mathbb{R}^m$ is a column vector such that the i-th element is one and the other elements are zeros, and $e_{Y,j} \in \mathbb{R}^n$ is a column vector such that the j-th element is one and the other elements are zeros.

Let $T \in B(H)$ be nonnegative definite and $\{e_i\}$ be an orthonormal basis of H. If

$$\|T\|_{TR} := \sum_{j=1}^{\infty} \langle Te_j, e_j \rangle$$

is finite, we say that $\|T\|_{TR}$ is the trace norm of T and that T is a trace class. Similar to an HS norm value, a trace norm value does not depend on the choice of orthonormal basis $\{e_j\}$.

If we substitute $x = e_j$ into (2.16) in Proposition 27, then we have $Tx = \lambda e_j$ and obtain that

$$\|T\|_{TR} := \sum_{j=1}^{\infty} \langle Te_j, e_j \rangle = \sum_{j=1}^{\infty} \lambda_j \, .$$

On the other hand, from

$$\|T\|_{HS}^2 = \sum_{i=1}^{\infty}\sum_{j=1}^{\infty} \langle Te_{i,1}, e_{j,2}\rangle^2 = \sum_{j=1}^{\infty} \lambda_j^2$$

we have

$$\|T\|_{HS} \leq \left(\lambda_1 \sum_{i=1}^{\infty} \lambda_i\right)^{1/2} = \sqrt{\lambda_1 \|T\|_{TR}} \ .$$

Thus, we have established the following proposition.

Proposition 31 *If $T \in B(H)$ is a trace class, it is a compact HS class.*

Appendix: Proofs of Propositions

Proof of Proposition 13

We show that a simple function approximates an arbitrary $f \in L_2^2$ and that a continuous function approximates an arbitrary simple function. Hereafter, we denote the L^2 norm by $\|\cdot\|$.

Since $f \in L_2$ is measurable, if f is nonnegative, the sequence $\{f_n\}$ of simple functions defined by

$$f_n(\omega) = \begin{cases} (k-1)2^{-n}, & (k-1)2^{-n} \leq f(\omega) < k2^{-n}, \ 1 \leq k \leq n2^n \\ n, & n \leq f(\omega) \leq \infty \end{cases}$$

satisfies $0 \leq f_1(\omega) \leq f_2(\omega) \leq \cdots \leq f(\omega)$ and $|f_n(\omega) - f(\omega)|^2 \to 0$ almost surely. Since the right-hand side of $|f_n(\omega) - f(\omega)|^2 \leq 4\{f(\omega)\}^2$ is finite when integrated, from the dominant convergence theorem, we have

$$\|f_n - f\|^2 \to 0 \ .$$

We can show a similar derivation for a general f that is not necessarily nonnegative, as derived in Chap. 1.

On the other hand, let A be a closed subset of $[a, b]$, and let K_A be the indicator function ($K_A(e) = 1$ if $e \in A$; $K_A(e) = 0$ otherwise). If we define $h(x) := \inf_{y \in A}\{|x - y|\}$ and $g_n^A(x) := \dfrac{1}{1 + nh(x)}$, then g_n^A is continuous, $g_n^A(x) \leq 1$ for $x \in [a, b]$, $g_n^A(x) = 1$ for $x \in A$, and $\lim_{n \to \infty} g_n^A(x) = 0$ for $x \in B := [a, b]\backslash A$. Thus, we have

$$\lim_{n\to\infty} \|g_n^A - K_A\| = \lim_{n\to\infty} \left(\int_B g_n^A(x)^2 dx \right)^{1/2} = \left(\int_B \lim_{n\to\infty} g_n^A(x)^2 dx \right)^{1/2} = 0 ,$$

where the second equality follows from the dominant convergence theorem. Moreover, if A, A' are disjoint, then $\alpha g_n^A + \alpha' g_n^{A'}$ with $\alpha, \alpha' > 0$ approximates $\alpha K_A + \alpha' K_{A'}$. In fact, we have

$$\|\alpha g_n^A + \alpha' g_n^{A'} - (\alpha K_A + \alpha' K_{A'})\| \le \alpha \|g_n^A - K_A\| + \alpha' \|g_n^{A'} - K_{A'}\| .$$

Hence, a sequence of continuous functions can approximate an arbitrary simple function.   □

Proof of Proposition 14

Suppose that $\{f_n\}$ is a Cauchy sequence in L^2, which means that

$$\lim_{N\to\infty} \sup_{m,n\ge N} \|f_m - f_n\|_2 = 0 . \tag{2.18}$$

Then, there exists a sequence $\{n_k\}$ such that

$$\left\| \sum_{k=1}^\infty |f_{n_{k+1}} - f_{n_k}| \right\|_2 \le \sum_{k=1}^\infty \|f_{n_{k+1}} - f_{n_k}\|_2 < \infty .$$

Thus, almost surely, we have

$$\sum_{k=1}^\infty |f_{n_{k+1}}(x) - f_{n_k}(x)| < \infty . \tag{2.19}$$

For arbitrary $r < t$ and $x \in E$, from the triangle inequality, we have

$$|f_{n_r}(x) - f_{n_t}(x)| \le \sum_{k=r}^{t-1} |f_{n_{k+1}}(x) - f_{n_k}(x)| .$$

Combined with (2.19), the real sequence $\{f_{n_k}(x)\}_{k=1}^\infty$ is almost surely Cauchy. Since the entire real system is complete (Proposition 6), we define $f(x) := \lim_{k\to\infty} f_{n_k}(x)$ for $x \in E$ such that $\{f_{n_k}(x)\}_{k=1}^\infty$ is Cauchy, and we define $f(x) := 0$ for the other $x \in E$. From (2.18), for an arbitrary $\epsilon > 0$, we have

$$\|f - f_n\|_2 = \int_E |f_n - f|^2 d\mu = \int_E \liminf_{k\to\infty} |f_n - f_{n_k}|^2 d\mu \le \liminf_{k\to\infty} \int_E |f_n - f_{n_k}|^2 d\mu < \epsilon$$

as $n \to \infty$, where the first inequality is due to Fatou's lemma. Furthermore, since $f_n, f - f_n \in L^2$ and L^2 is a linear space, we have $f \in L_2$.   □

Proof of Proposition 15

The first item holds because

$$0 \leq \|x - \sum_{i=1}^{n} \langle x, e_i \rangle e_i\|^2 = \|x\|^2 - \sum_{i=1}^{n} \langle x, e_i \rangle^2$$

for all n. For the second item, letting $n > m$, $s_n := \sum_{k=1}^{n} \langle x, e_k \rangle e_k$, we have

$$\|s_n - s_m\|^2 = \langle \sum_{k=m+1}^{n} \langle x, e_k \rangle e_k, \sum_{k=m+1}^{n} \langle x, e_k \rangle e_k \rangle = \sum_{k=m+1}^{n} |\langle x, e_k \rangle|^2,$$

which diminishes as $n, m \to \infty$ according to the first item. For the third item, we have

$$\|s_n - s_m\|^2 = \langle \sum_{k=m+1}^{n} \alpha_k e_k, \sum_{k=m+1}^{n} \alpha_k e_k \rangle = \sum_{k=m+1}^{n} \alpha_k^2 = S_n - S_m$$

for $s_n := \sum_{i=1}^{n} \alpha_i e_i$, $S_n := \sum_{i=1}^{n} \alpha_i^2$, and $n > m$. Thus, the third item follows from the equivalence: $\{s_n\}$ is Cauchy $\iff \{S_n\}$ is Cauchy.

The last item holds because $\langle y, e_i \rangle = \lim_{n \to \infty} \langle \sum_{j=1}^{n} \alpha_j e_j, e_i \rangle = \alpha_i$ for $y = \sum_{j=1}^{\infty} \alpha_j e_j$, which follows from the continuity of inner products (Proposition 9). $\qquad \square$

Proof of Proposition 16

For 1.$\Rightarrow$6., since $\{e_i\}$ is an orthonormal basis of H, we may write an arbitrary $x \in H$ as $x = \sum_{i=1}^{\infty} \alpha_i e_i$, $\alpha_i \in \mathbb{R}$. From the fourth item of Proposition 15, we have $\alpha_i = \langle x, e_i \rangle$ and obtain 6. 6.$\Rightarrow$5. is obtained by substituting $x = \sum_{i=1}^{\infty} \langle x, e_i \rangle e_i$, $y = \sum_{i=1}^{\infty} \langle y, e_i \rangle e_i$ into $\langle x, y \rangle$. 5.$\Rightarrow$4. is obtained by substituting $x = y$ in 5. 4.$\Rightarrow$3. is due to

$$\|x - \sum_{k=1}^{n} \langle x, e_k \rangle e_k\|^2 = \|x\|^2 - \sum_{k=1}^{n} |\langle x, e_k \rangle|^2 \to 0$$

as $n \to \infty$ for each $x \in H$. For 3.$\Rightarrow$2., note the implication $\langle x, e_k \rangle = 0, k = 1, 2, \ldots$ $\Rightarrow x \perp \text{span}\{e_k\}$, which implies that $x \perp \overline{\text{span}\{e_k\}}$ from the continuity of inner products (Proposition 9). Thus, we have $\langle x, x \rangle = 0$ and $x = 0$. For 2.$\Rightarrow$1., from the second item of Proposition 15, $y = \sum_{i=1}^{\infty} \langle z, e_i \rangle e_i$ converges for each $z \in H$. Therefore, for each j, we have

$$\langle z - y, e_j \rangle = \langle z, e_j \rangle - \lim_{n \to \infty} \langle \sum_{i=1}^{n} \langle z, e_i \rangle, e_j \rangle = \langle z.e_j \rangle - \langle z.e_j \rangle = 0 \,.$$

From the assumption of 2, we have that $z - y = 0$ and that z can be written as $\sum_{i=1}^{\infty} \langle z, e_i \rangle e_i$.   □

Proof of Proposition 19

Let M be a closed subset of H. We show that for each $x \in H$, there exists a unique $y \in M$ that minimizes $\|x - y\|$ and that we have

$$\langle x - y, z - y \rangle \leq 0 \tag{2.20}$$

for $z \in M$. To this end, we first show that any sequence $\{y_n\}$ in M for which

$$\lim_{n \to \infty} \|x - y_n\|^2 = \inf_{y \in M} \|x - y\|^2 \tag{2.21}$$

is Cauchy. Since M is a linear space, we have $(y_n + y_m)/2 \in M$ and

$$\|y_n - y_m\|^2 = 2\|x - y_n\|^2 + 2\|x - y_m\|^2 - 4\|x - \frac{y_n + y_m}{2}\|^2$$
$$\leq 2\|x - y_n\|^2 + 2\|x - y_m\|^2 - 4 \inf_{y \in M} \|x - y\|^2 \to 0 \ .$$

Hence, $\{y_n\}$ is Cauchy. Then, suppose that more than one lower limit y exists, and let $u \neq v$ be such a y. For example, for $\{y_n\}$, let $y_{2m-1} \to u$, and let $y_{2m} \to v$ satisfy (2.21). However, this limit is not Cauchy and contradicts the discussion shown thus far. Hence, the y that achieves the limit in (2.21) is unique. In the following, we assume that y gives the lower limit.

Moreover, note that

$$\|x - \{az + (1-a)y\}\|^2 \geq \|x - y\|^2 \iff 2a\langle x - y, z - y \rangle \leq a^2\|z - y\|^2$$

for arbitrary $0 < a < 1$ and $z \in M$, and if $\langle x - y, z - y \rangle > 0$, the inequality flips for small $a > 0$. Thus, we have $\langle x - y, z - y \rangle \leq 0$.

Finally, if we substitute $z = 0, 2y$ into (2.20), we have $\langle x - y, y \rangle = 0$. Therefore, (2.20) implies that $\langle x - y, z \rangle \leq 0$ for $z \in M$. We obtain the proposition by replacing z with $-z$.   □

Proof of Proposition 22

If the operator T maps to zero for any element, then $e_T = 0$ satisfies the desired condition. Thus, we assume that T outputs a nonzero value for at least one input. From the first item of Proposition 20, $\text{Ker}(T)^\perp$ is a closed subset of H and contains a y such that $Ty = 1$. Thus, for an arbitrary $x \in H$, we have

$$T(x - (Tx)y) = Tx - TxTy = 0$$

and $x - (Tx)y \in \text{Ker}(T)$. Since $y \in \text{Ker}(T)^\perp$, we have $\langle x - (Tx)y, y \rangle = 0$ and

$$\langle x, y \rangle = Tx \langle y, y \rangle = Tx \|y\|^2 .$$

Thus, $e_T = y/\|y\|^2$ satisfies the desired condition.

To demonstrate uniqueness, if e'_T satisfies the same condition, then $\langle x, e_T - e'_T \rangle = 0$ for any $x \in H$, which means that $e_T = e'_T$.

Furthermore, since $\|Tx\| = \langle x, e_T \rangle \le \|x\| \|e_T\|$ for $x \in H$, we have that $\|T\| \le \|e_T\|$ when $\|x\| = 1$. Additionally, we obtain the inverse inequality $\|e_T\| = \frac{1}{\|y\|} = \frac{\|Ty\|}{\|y\|} \le \|T\|$.   □

Proof of Proposition 25

For the first item, note that if $\{x_n\}$ is bounded, so is $\{Tx_n\}$. Moreover, if the image of T is of finite dimensionality, then $\{Tx_n\}$ is also compact (Proposition 7)[9]. For the second item, we use the so-called diagonal argument. In the following, we denote the norms of H_1, H_2 by $\|\cdot\|_1$, $\|\cdot\|_2$. Let $\{x_k\}$ be a bounded sequence in X_1. From the compactness of T_1, there exists $\{x_{1,k}\} \subseteq \{x_{0,k}\} := \{x_k\}$ such that $\{T_1 x_{1,k}\}$ converges to a $y_1 \in H_2$ as $k \to \infty$. Then, there exists $\{x_{2,k}\} \subseteq \{x_{1,k}\}$ such that $\{T_2 x_{2,k}\}$ converges to a $y_2 \in H_2$ as $k \to \infty$. If we repeat this process, the sequence $\{y_n\}$ in H_2 converges. In fact, for each n, there exists a large k_n such that

$$\|T_n x_{n,k} - y_n\|_2 < \frac{1}{n}, \ k \ge k_n .$$

If we make $\{k_n\}$ monotone, then for $m < n$, we obtain

$$y_m - y_n = (y_m - T_m x_{n,k_n}) + (T_n x_{n,k_n} - y_n) + (T_m x_{n,k_n} - T x_{n,k_n}) + (T x_{n,k_n} - T_n x_{n,k_n}) .$$

Thus, as $m, n \to \infty$, we have

$$\|y_m - y_n\| \le \frac{1}{m} + \frac{1}{n} + \|T_m - T\| \cdot \|x_{n,k_n}\|_1 + \|T_n - T\| \cdot \|x_{n,k_n}\|_1 \to 0.$$

Since H_2 is complete, there exists a $y \in H_2$ such that $\{y_n\}$ converges. Since

$$\|T x_{n,k_n} - y\|_2 \le \|T - T_n\| \|x_{n,k_n}\|_1 + \|T_n x_{n,k_n} - y_n\|_2 + \|y_n - y\|_2 \to 0$$

as $n \to \infty$, we have shown that $\{Tx_n\}$ has a convergent subsequence in H_2.   □

Proof of Proposition 26

By induction, we show that

---

[9] This statement is called Bolzano-Weierstrass's theorem for sequential compactness rather than Heine-Borel's theorem. The two theorems coincide for metric spaces.

$$\sum_{j=1}^{n} c_j e_j = 0 \Rightarrow c_1 = c_2 = \cdots = c_n = 0 . \tag{2.22}$$

For $n = 2$, suppose that $c_1 e_1 + c_2 e_2 = 0$. Then, we have $T(c_1 e_1 + c_2 e_2) = \lambda_1 c_1 e_1 + \lambda_2 c_2 e_2 = 0$. From these two equations and $\lambda_1 \neq \lambda_2$, we have $c_1 = c_2 = 0$. Thus, we obtain (2.22) for $n = 2$. For $n = k$, $\sum_{j=1}^{k+1} c_j e_j = 0$ and $\sum_{j=1}^{k+1} \lambda_j c_j e_j = 0$ imply that

$$0 = \lambda_{k+1} \sum_{j=1}^{k+1} c_j e_j - \sum_{j=1}^{k+1} \lambda_j c_j e_j = \sum_{j=1}^{k} (\lambda_{k+1} - \lambda_j) c_j e_j .$$

From $\lambda_{k+1} \neq \lambda_j$, if we assume that $c'_j := (\lambda_{k+1} - \lambda_j) c_j \neq 0$, then from $\sum_{j=1}^{k} c'_j e_j = 0$ and the assumption of induction, we have $c'_1 = \cdots = c'_k = 0$, which means that $c_1 = \cdots = c_k = 0$ and $c_{k+1} e_{k+1} = -\sum_{j=1}^{k} c_j e_j = 0$. Thus $c_{k+1} = 0$. Moreover, under the condition that T is self-adjoint, from $\langle e_i, e_j \rangle = \langle e_i, \lambda_j^{-1} T e_j \rangle = \lambda_j^{-1} \langle T e_i, e_j \rangle = \lambda_j^{-1} \lambda_i \langle e_i, e_j \rangle$ and $\lambda_i \neq \lambda_j$ for $i \neq j$, we have $\langle e_i, e_j \rangle = 0$. Thus, the $\{e_j\}$ are orthogonal. $\qquad\square$

Proof of Proposition 27

We first show that

$$\mathrm{Ker}(T)^{\perp} = \overline{\mathrm{Im}(T^*)} . \tag{2.23}$$

For $x_1 \in \mathrm{Ker}(T)$ and $x_2 \in H$, we see that $\langle x_1, T^* x_2 \rangle_1 = \langle T x_1, x_2 \rangle_2 = 0$ and that x_1 is orthogonal to any element of $\mathrm{Im}(T^*)$. Thus, we have

$$\mathrm{Ker}(T) \subseteq (\mathrm{Im}(T^*))^{\perp}.$$

Moreover, if $x_1 \in (\mathrm{Im}(T^*))^{\perp}$, from $T^*(T x_1) \in \mathrm{Im}(T^*)$, we have

$$\|T x_1\|_2 = \langle x_1, T^* T x_1 \rangle_1 = 0 ,$$

which means that $x_1 \in \mathrm{Ker}(T)$ and establishes inverse inclusion. Thus, we have shown that $\mathrm{Ker}(T) = (\mathrm{Im}(T^*))^{\perp}$. Furthermore, if we apply the third item of Proposition 20, we obtain

$$(\mathrm{Ker}(T))^{\perp} = \overline{\mathrm{Im}(T^*)} .$$

Note that since $\mathrm{Ker}(T)$ is an orthogonal complement of subset $\mathrm{Im}(T^*)$ of H, the first item of Proposition 20 and (2.11) can be applied. Since $T \in B(H)$ is self-adjoint ($T^* = T$), we can write (2.23) further as

$$H = \mathrm{Ker}(T) \oplus \overline{\mathrm{Im}(T)} .$$

Hence, in order to show that (2.16), it is sufficient to prove that

$$\overline{\text{Im}(T)} = \overline{\text{span}\{e_j : j \geq 1\}} . \tag{2.24}$$

Note that for each finite $n = 1, 2, \ldots$ and $c_1, c_2, \ldots, c_n \in \mathbb{R}$, we have

$$\sum_{j=1}^{n} c_j e_j = T \left(\sum_{j=1}^{n} \lambda_j^{-1} c_j e_j \right)$$

and $\text{span}\{e_j | j \geq 1\} \subseteq \text{Im}(T)$. Even if we perform closure on both sides, the inclusion relation does not change. Thus, we have $\overline{\text{span}\{e_j | j \geq 1\}} \subseteq \overline{\text{Im}(T)}$. Furthermore, we decompose (2.11)

$$\overline{\text{Im}(T)} = \overline{\text{span}\{e_j | j \geq 1\}} \oplus N ,$$

where $N = \overline{\text{span}\{e_j | j \geq 1\}}^{\perp} \cap \overline{\text{Im}(T)}$. Note that $Ty \in \overline{\text{span}\{e_j | j \geq 1\}}$ for $y \in \text{span}\{e_j | j \geq 1\}$, and

$$\langle Tx, y \rangle = \langle x, Ty \rangle = 0$$

for $x \in N$ because T is self-adjoint. Thus, we have $Tx \in N$.

Now, in general, we have

$$\|T\| = w(T) := \sup_{\|x\|=1} |\langle Tx, x \rangle| . \tag{2.25}$$

In fact,

$$|\langle Tx, y \rangle| = \left| \frac{1}{4} \langle T(x+y), x+y \rangle - \frac{1}{4} \langle T(x-y), x-y \rangle \right|$$

$$\leq \frac{1}{4} |\langle T(x+y), x+y \rangle| + \left| \frac{1}{4} \langle T(x-y), x-y \rangle \right|$$

$$\leq \frac{1}{4} (w(T)(\|x+y\|^2 + \|x-y\|^2) = \frac{1}{2} w(T)(\|x\|^2 + \|y\|^2),$$

and if we take the upper limit under $\|x\| = \|y\| = 1$, we obtain

$$\|T\| = \sup_{\|x\|=1} \langle Tx, \frac{Tx}{\|Tx\|} \rangle \leq \sup_{\|x\|=\|y\|=1} \langle Tx, y \rangle \leq w(T) .$$

On the other hand, we have

$$w(T) \leq \sup_{\|x\|=1} \|Tx\| \cdot \|x\| = \sup_{\|x\|=1} \|Tx\| = \|T\|$$

and (2.25).

In addition, we know that either $\pm \|T\|$ is an eigenvalue of T. In fact, from (2.25), there exists a sequence $\{x_n\}$ in H with $\|x_n\| = 1$ such that $\langle Tx_n, x_n \rangle \to \|T\|$ or $\langle Tx_n, x_n \rangle \to -\|T\|$ (the upper and lower limits are convergence points). For the former case, we have

$$0 \le \|Tx_n - \|T\| \, x_n\|^2 = \|Tx_n\|^2 + \|T\|^2 \|x_n\|^2 - 2\|T\|\langle Tx_n, x_n \rangle \to 0.$$

From compactness of T, there exists $\{x_{n_k}\}$ ($\subseteq \{x_n\}$) such that $Tx_{n_k} \to y \in H$. From $Tx_{n_k} - \|T\| \, x_{n_k} \to 0$, there exists $0 \ne x \in H$ such that $\|T\|x_{n_k} \to \|T\|x$. From $\|T\| \, x = y = \lim_{k\to\infty} Tx_{n_k}$, we have that $Tx = \|T\|x$ and that $\|T\|$ is an eigenvalue of T. For the latter case, $-\|T\|$ is an eigenvalue of T.

Finally, we assume that there exists an $x \in N$ such that $\|Tx\| \ne 0$. Let T_N be the restriction of T on N. Because $\|T_N\| > 0$, either $\|T_N\|$ or $-\|T_N\|$ is an eigenvalue of T. The existence of an eigenvalue on N contradicts the chosen orthonormal basis $\{e_j\}_{j=1}^{\infty}$. Therefore, when $x \in N$, we have $Tx = 0$, which means that $N \subseteq \overline{\mathrm{Im}(T)} \cap \mathrm{Ker}(T) = \{0\}$. Thus, we have established (2.16).

$\square$

Exercises 16~30

16. Choose the closed sets among the sets below. For the nonclosed sets, find their closures.

    (a) $\bigcup_{n=1}^{\infty} [n - \frac{1}{n}, n + \frac{1}{n}]$;
    (b) $\{2, 3, 5, 7, 11, 13, \ldots\}$;
    (c) $\mathbb{R} \cap \mathbb{Z}$;
    (d) $\{(x, y) \in \mathbb{R}^2 \mid x^2 + y^2 < 1 \text{ when } x \ge 0, \ x^2 + y^2 \le 1 \text{ when } x < 0\}$.

17. Show that the sequence $a_1 = 1,\ a_{n+1} = \dfrac{1}{2}a_n + \dfrac{1}{a_n}$ converges to $\sqrt{2}$ as $n \to \infty$.

18. Let $f : M \to \mathbb{R}$ be a function defined over a bounded closed set M, and we define $\Delta(z_1), \ldots, \Delta(z_m)$ for some $m \ge 1$ and $z_1, \ldots, z_m$ such that

$$d(x, z) < \Delta(z) \Longrightarrow d(f(x), f(z)) < \epsilon$$

for $z \in M$.

(a) Why can the neighborhoods cover M ?

Let $x, y \in M$ satisfy $d_1(x, y) < \delta := \dfrac{1}{2} \min_{1 \le i \le m} \Delta(z_i)$. Without loss of generality, we assume that $x \in U_i$ with a center at z_i and a radius of $\Delta(z_i)/2$. Prove the following:

(b) $d_1(x, z_i) < \frac{1}{2}\Delta(z_i) < \Delta(z_i)$.
(c) $d_1(y, z_i) \le d_1(x, y) + d_1(x, z_i) < \Delta(z_i)$.
(d) $d_2(f(x), f(y)) \le d_2(f(x), f(z_i)) + d_2(f(y), f(z_i)) < \epsilon + \epsilon = 2\epsilon$.
(e) f is uniformly continuous.

19. Using the fact that any continuous function over a bounded closed set is uniformly continuous, show that a continuous function over $[0, 1]$ is a Riemann integral.

20. Show that the Cauchy-Schwartz inequality (2.5) holds if and only if one of x, y is a constant multiplied by the other.

21. Show that a one-indeterminate polynomial ring A is an algebra. In addition, show that the set of functions $f \in A$ over $E := [0, 1]$ is dense in $C(E)$.

22. Derive Riesz-Fischer's theorem stating that "L^2 is complete" (Proposition 14) according to the following steps in the appendix:

    (a) Let $\{f_n\}$ be an arbitrary Cauchy sequence.
    (b) There exists a sequence $\{n_k\}$ such that $\| \sum_{k=1}^{\infty} | f_{n_{k+1}} - f_{n_k} | \|_2 < \infty$.
    (c) Prove the existence of an $f : E \to \mathbb{R}$ such that $\mu\{x \in E \,|\, \lim_{k \to \infty} f_{n_k}(x) = f(x)\} = \mu(E)$.
    (d) Show that $\| f_n - f \| \to 0$ and $f \in L^2[a, b]$.

23. Show that the basis of the Fourier series expansion

$$\left\{ \frac{1}{\sqrt{2\pi}}, \frac{\cos x}{\sqrt{\pi}}, \frac{\sin x}{\sqrt{\pi}}, \frac{\cos 2x}{\sqrt{\pi}}, \frac{\sin 2x}{\sqrt{\pi}}, \cdots \right\}.$$

is orthonormal.

24. Derive Proposition 19 according to the following steps in the appendix. What are the derivations of (a) through (e)?

    (a) Show that a sequence $\{y_n\}$ in M for which

$$\lim_{n \to \infty} \| x - y_n \|^2 = \inf_{y \in M} \| x - y \|^2$$

    converges in M. Hereafter, let y satisfy $y_n \to y \in M$.
    (b) Show that $2a \langle x - y, z - y \rangle \le a^2 \| z - y \|^2$ for $0 < a < 1$ and $z \in M$.
    (c) Show that the inequality $\langle x - y, z - y \rangle > 0$ contains a contradiction.
    (d) Show that $\langle x - y, z \rangle \le 0$.
    (e) Obtain the proposition by replacing z with $-z$.

25. Show that the linear operator norm (2.12) satisfies the triangle inequality.

26. Show that the integral operator (2.13) is a bounded linear operator and that it is self-adjoint when K is symmetric.

27. Let (M, d) be a metric space with $M := \mathbb{R}$ and a Euclidean distance d. Show that each of the following $E \subseteq M$ is not sequentially compact. Furthermore, show that they are not compact without using the equivalence between compactness and sequential compactness.

    (a) $E = [0, 1)$;
    (b) $E = \mathbb{Q}$.

28. Proposition 27 is derived according to the following steps in the appendix. What are the derivations of (a) through (c)?

(a) Show that $H_1 = \mathrm{Ker}(T) \oplus \overline{\mathrm{Im}(T)}$.
(b) Show that $\overline{\mathrm{span}\{e_j | j \geq 1\}} \subseteq \overline{\mathrm{Im}(T)}$.
(c) Show that $\overline{\mathrm{span}\{e_j | j \geq 1\}} \supseteq \overline{\mathrm{Im}(T)}$.

Why do we need to show (2.25)?

29. Show that the HS and trace norms satisfy the triangle inequality.
30. Show that if $T \in B(H)$ is a trace class, then it is also an HS class, and show that if $T \in B(H)$ is a trace class, it is also compact.

Chapter 3
Reproducing Kernel Hilbert Space

Thus far, we have learned that a feature map $\Psi : E \ni x \mapsto k(x, \cdot)$ is obtained by the positive definite kernel $k : E \times E \to \mathbb{R}$. In this chapter, we generate a linear space H_0 based on its image $k(x, \cdot)(x \in E)$ and construct a Hilbert space H by completing this linear space, where H is called the reproducing kernel Hilbert space (RKHS), which satisfies the reproducing property of the kernel k (k is the reproducing kernel of H). In this chapter, we first understand that there is a one-to-one correspondence between the kernel k and the RKHS H and that H_0 is dense in H (via the Moore-Aronszajn theorem). Furthermore, we introduce the RKHS represented by the sum of RKHSs and apply it to Sobolev spaces. We prove Mercer's theorem regarding integral operators in the second half of this chapter and compute their eigenvalues and eigenfunctions. This chapter is the core of the theory contained in this book, and the later chapters correspond to its applications.

3.1 RKHSs

Let H be a Hilbert space whose elements are functions $f : E \to \mathbb{R}$.

A function $k : E \times E \to \mathbb{R}$ is said to be a reproducing kernel of a Hilbert space H with an inner product $\langle \cdot, \cdot \rangle_H$ if it satisfies the following two conditions:

1. For each $x \in E$, we have
$$k(x, \cdot) \in H. \tag{3.1}$$

2. Reproducing property: for each $f \in H$ and $x \in E$,
$$f(x) = \langle f, k(x, \cdot) \rangle_H. \tag{3.2}$$

© The Author(s), under exclusive license to Springer Nature Singapore Pte Ltd. 2022
J. Suzuki, *Kernel Methods for Machine Learning with Math and R*,
https://doi.org/10.1007/978-981-19-0398-4_3

When H has a reproducing kernel, we say that H is a reproducing kernel Hilbert space (RKHS). The reproducing property (3.2) is called a kernel trick.

Example 51 Let $\{e_1, \ldots, e_p\}$ be an orthonormal basis of a finite-dimensional Hilbert space H. If we define

$$k(x, y) := \sum_{i=1}^{p} e_i(x) e_i(y) \tag{3.3}$$

for $x, y \in E$, then we have $k(x, \cdot) \in H$ and

$$\langle e_j(\cdot), k(x, \cdot) \rangle_H = \sum_{i=1}^{p} \langle e_j, e_i \rangle_H e_i(x) = e_j(x)$$

for each $1 \le j \le p$. Thus, for any $f(\cdot) = \sum_{i=1}^{p} f_i e_i(\cdot) \in H$, $f_i \in \mathbb{R}$, we have $\langle f(\cdot), k(x, \cdot) \rangle_H = f(x)$ (reproducing property). Therefore, H is an RKHS, and (3.3) is a reproducing kernel.

Proposition 32 *The reproducing kernel k of the RKHS H is unique, symmetric $k(x, y) = k(y, x)$, and nonnegative definite.*

Proof. If k_1, k_2 are RKHSs of H, then by the reproducing property, we have that

$$f(x) = \langle f, k_1(x, \cdot) \rangle_H = \langle f, k_2(x, \cdot) \rangle_H .$$

In other words,

$$\langle f, k_1(x, \cdot) - k_2(x, \cdot) \rangle_H = 0$$

holds for all $f \in H$, $x \in E$ for which $k_1 = k_2$ (Proposition 16). Additionally, the symmetry of a reproducing kernel follows from that of its inner product:

$$k(x, y) = \langle k(x, \cdot), k(y, \cdot) \rangle_H = \langle k(y, \cdot), k(x, \cdot) \rangle_H = k(y, x) .$$

The nonnegative definiteness of the reproducing kernel can be shown as follows:

$$\sum_{i=1}^{n} \sum_{j=1}^{n} z_i z_j k(x_i, x_j) = \sum_{i=1}^{n} \sum_{j=1}^{n} z_i z_j \langle k(x_i, \cdot), k(x_j, \cdot) \rangle_H = \langle \sum_{i=1}^{n} z_i k(x_i, \cdot), \sum_{j=1}^{n} z_j k(x_j, \cdot) \rangle_H \ge 0.$$

$\square$

Proposition 33 *A Hilbert space H is an RKHS if and only if $T_x(f) = f(x)$ ($f \in H$) is bounded at each $x \in E$ for the linear functional $T_x : H \ni f \mapsto f(x) \in \mathbb{R}$.*

Proof. If H has a reproducing kernel k, then at each $x \in E$, we have

$$\langle f(\cdot), k(x, \cdot) \rangle_H = T_x(f) , \quad f \in H .$$

Thus, we have

$$|T_x(f)| = |\langle f(\cdot), k(x, \cdot)\rangle_H| \le \|f\| \cdot \|k(x, \cdot)\| = \|f\|\sqrt{k(x, x)} .$$

Conversely, if the linear functional $T_x(f) = f(x)$ is bounded for $x \in E$, from Proposition 22, there exists a $k_x : E \to \mathbb{R}$ such that

$$\langle f(\cdot), k_x(\cdot)\rangle_H = f(x) , \quad f \in H .$$

In other words, a reproducing kernel exists.   □

In Proposition 32, we showed that a reproducing kernel is unique once its RKHS is determined, but the following proposition asserts the converse.

Proposition 34 (Aronszajn [1]) *Let $k : E \times E \to \mathbb{R}$ be a positive definite kernel. Then, the Hilbert space H with the reproducing kernel k is unique. Moreover, for $k(x, \cdot) \in H$, $x \in E$ holds, and the generated linear space is dense in H.*

The proof is given by the following procedure:

1. Define the inner product $\langle \cdot, \cdot \rangle_{H_0}$ of $H_0 := \text{span}\{k(x, \cdot)|x \in E\}$.
2. For any Cauchy sequence $\{f_n\}$ in H_0 and each $x \in E$, the real sequence $\{f_n(x)\}$ is a Cauchy sequence, and we have the convergence value $f(x) := \lim_{n\to\infty} f_n(x)$ (Proposition 6). Let H be such a set of f.
3. Define the inner product $\langle \cdot, \cdot \rangle_H$ of the linear space H.
4. Show that H_0 is dense in H.
5. Show that any Cauchy sequence $\{f_n\}$ in H converges to some element of H as $n \to \infty$ (completeness of H).
6. Show that k is a reproducing kernel of H.
7. Show that such an H is unique.

See the appendix at the end of the chapter for details.[1]   □

Example 52 (*Linear Kernel*) Let $\langle \cdot, \cdot \rangle_E$ be the inner product of $E := \mathbb{R}^d$. Then, the linear space

$$H := \{\langle x, \cdot \rangle_E | x \in E\}$$

is complete since it has finite dimensions (Proposition 6). Moreover, H is an RKHS with the reproducing kernel $k(x, y) = \langle x, y \rangle_E$, $x, y \in E$.

Example 53 Let E be a finite set $\{x_1, \ldots, x_n\}$, and let $k : E \times E \to \mathbb{R}$ be a positive definite kernel, then the linear space

$$H := \{\sum_{i=1}^{n} \alpha_i k(x_i, \cdot)|\alpha_1, \ldots, \alpha_n \in \mathbb{R}\}$$

---

[1] The proof is due to [33].

is a reproducing kernel Hilbert space. We define the inner product by

$$\langle f(\cdot), g(\cdot) \rangle_H = a^\top K b$$

for $f(\cdot), g(\cdot) \in H$, where $f(\cdot) = \sum_{j=1}^n a_j k(x_j, \cdot) \in H$, $a = [a_1, \ldots, a_n]^\top \in \mathbb{R}^n$ and $g(\cdot) = \sum_{j=1}^n b_j k(x_j, \cdot) \in H$, $b = [b_1, \ldots, b_n]^\top \in \mathbb{R}^n$ via the Gram matrix

$$K := \begin{bmatrix} k(x_1, x_1) & \cdots & k(x_1, x_n) \\ \vdots & \ddots & \vdots \\ k(x_n, x_1) & \cdots & k(x_n, x_n) \end{bmatrix}.$$

Then, for each x_i, $i = 1, 2, \ldots$, we have

$$\langle f(\cdot), k(x_i, \cdot) \rangle_H = [a_1, \ldots, a_n] K e_i = \sum_{j=1}^n a_j k(x_j, x_i) = f(x_i)$$

(reproducing property), where e_i is an n-dimensional column vector in which we set component i and the other components to 1 and 0, respectively.

Example 54 (*Polynomial Kernel*) Let $\langle \cdot, \cdot \rangle_E$ be the inner product between the elements in E. The Hilbert space H obtained by completing the linear space H_0 generated by $(\langle x, \cdot \rangle_E + 1)^d \in \mathbb{R}$ ($x \in E$) is an RKHS with the reproducing kernel $k(x, y) = (\langle x, y \rangle_E + 1)^d$ for $x, y \in E$.

Example 55 Let $k(x, y)$ be the kernel expressed by a function $\phi(x - y)$ as considered in Section 1.5. If we require $k(x, y)$ to take real values, the associated probability density functions must be even functions such as those of the Gaussian and Laplace distributions. Otherwise, since the imaginary part of $t \mapsto e^{i(x-y)t}$ is odd, the kernel k might take imaginary values. Now, using $L^2(E, \eta) \ni F : E = \mathbb{R} \to \mathbb{C}$ whose real and imaginary parts are even and odd, respectively, we consider the linear space consisting of $f : E \to \mathbb{R}$ with $f(x) = \int_E F(t) e^{ixt} d\eta(t)$. The function $F(t) \mapsto f(x) = \int_E F(t) e^{ixt} d\eta(t)$ is injective (if $\int_E F(t) e^{ixt} d\eta(t) = 0$, then the inverse Fourier transform $F(t) = 0$). If its inner product is $\langle f, g \rangle_H = \int_E F(t) \overline{G(t)} d\eta(t)$ for $F, G \in L^2(E, \eta)$, then $L^2(E, \eta)$ and

$$H = \left\{ E \ni x \mapsto \int_E F(t) e^{ixt} d\eta(t) \in \mathbb{R} \mid F \in L^2(E, \eta) \right\}$$

are isomorphic as an inner product space. Note that H has a reproducing kernel $E \times E \to \mathbb{R}$ with

$$k(x, y) = \int_E e^{-i(x-y)t} d\eta(t) .$$

In fact, we have $k(x, y) = \int_E e^{-ixt} e^{iyt} d\eta(t)$. Thus, if we set $G(t) = e^{-ixt}$, we obtain

$$\langle f(\cdot), k(x, \cdot) \rangle_H = \int_E F(t)\overline{G(t)}d\eta(t) = \int_E F(t)e^{ixt}d\eta(t) = f(x)$$

for $f(y) = \int_E F(t)e^{iyt}d\eta(t)$ and $k(x, y) = \int_E G(t)e^{iyt}d\eta(t)$. For different kernels $k(x, y)$, such as the Gaussian and Laplacian kernels, the measure $\eta(t)$ will be different, and the corresponding RKHS H will be different.

Example 56 Let $E := [0, 1]$. Using the real-valued function F with $\int_0^1 F(u)^2 du < \infty$, we consider the set H of functions $f : E \to \mathbb{R}$, $f(t) = \int_0^1 F(u)(t - u)_+^0 du$, where we denote $(z)_+^0 = 1$ and $(z)_+^0 = 0$ when $z \geq 0$ and when $z < 0$, respectively. The linear space H is complete for the norm $\|f\|^2 = \int_0^1 F(u)^2 du$ (Proposition 14) if the inner product is $\langle f, g \rangle_H = \int_0^1 F(u)G(u)du$ for $f(t) = \int_0^1 F(u)(t - u)_+^0 du$ and $g(t) = \int_0^1 G(u)(t - u)_+^0 du$. This Hilbert space H is the RKHS for $k(x, y) = \min\{x, y\}$. In fact, for each $z \in E$, we see that

$$\langle f(z), k(x, z) \rangle_H = \langle \int_0^1 F(u)(z - u)_+^0 du, \int_0^1 (x - u)_+^0 (z - u)_+^0 du \rangle_H = \int_0^1 F(u)(x - u)_+^0 du = f(x).$$

Thus far, we have obtained the RKHS corresponding to each positive definite kernel, but a necessary condition exists for a Hilbert space H to be an RKHS. If that condition is not satisfied, we can claim that it is not an RKHS.

Proposition 35 Let H be an RKHS consisting of functions on E. If $\lim_{n\to\infty} \|f_n - f\|_H = 0$ $f, f_1, f_2, \ldots \in H$, then for each $x \in E$, $\lim_{n\to\infty} |f_n(x) - f(x)| = 0$ holds.

Proof. In fact, we have that for each $x \in E$,

$$|f_n(x) - f(x)| \leq \|f_n - f\|\sqrt{k(x, x)}.$$

$\square$

Example 57 $H := L^2[0, 1]$ is not an RKHS. In fact, for a sequence $\{f_n\}$ with $f_n(x) = x^n$, the norm converges to $\|f_n\|_H^2 = \int_0^1 f_n^2(x)dx = \frac{1}{2n+1} \to 0$. However, for $f(x) = 0$ with $x \in E$, we have $\|f_n - f\|_H \to 0$, and $|f_n(1) - f(1)| = 1 \nrightarrow 0$. This contradicts the fact that H is an RKHS (Proposition 35).

Example 57 illustrates that $L^2[0, 1]$ is too large, and as we will see in the next section, the Sobolev space restricted to $L^2[0, 1]$ is an RKHS.

3.2 Sobolev Space

We first show that if k_1, k_2 are reproducing kernels, the sum $k_1 + k_2$ is also a reproducing kernel. To this end, we show the following.

Proposition 36 *If H_1, H_2 are Hilbert spaces, so is the direct product $F := H_1 \times H_2$ under the inner product*

$$\langle (f_1, f_2), (g_1, g_2) \rangle_F := \langle f_1, g_1 \rangle_{H_1} + \langle f_2, g_2 \rangle_{H_2} \tag{3.4}$$

for $f_1, g_1 \in H_1$, $f_2, g_2 \in H_2$.

Proof. From $\|(f_1, f_2)\|_F^2 = \|f_1\|_{H_1}^2 + \|f_2\|_{H_2}^2$, we have

$$\|f_{1,n} - f_{1,m}\|_{H_1}, \|f_{2,n} - f_{2,m}\|_{H_2}$$
$$\leq \sqrt{\|f_{1,n} - f_{1,m}\|_{H_1}^2 + \|f_{2,n} - f_{2,n}\|_{H_2}^2} = \|(f_{1,n}, f_{2,n}) - (f_{1,m}, f_{2,m})\|_F .$$

Thus, we have

$$\{(f_{1,n}, f_{2,n})\} \text{ is Cauchy}$$
$$\Rightarrow \{f_{1,n}\}, \{f_{2,n}\} \text{ is Cauchy}$$
$$\Rightarrow f_1 \in H_1, f_2 \in H_2 \text{ exists such that } f_{1,n} \to f_1, f_{2,n} \to f_2$$
$$\Rightarrow \|(f_{1,n}, f_{2,n}) - (f_1, f_2)\|_F = \|(f_{1,n} - f_1, f_{2,n} - f_2)\|_F$$
$$= \sqrt{\|f_{1,n} - f_1\|^2 + \|f_{2,n} - f_2\|^2} \to 0 ,$$

which means that F is complete.   □

Let

$$H := H_1 + H_2 := \{f_1 + f_2 | f_1 \in H_1, f_2 \in H_2\}$$

be the direct sum of H_1, H_2, and define the linear map from F to H by $u : F \ni (f_1, f_2) \mapsto f_1 + f_2 \in H$. Then, we can decompose F into $N := u^{-1}(0)$ and its orthogonal complement $N^\perp$. If we restrict u to $N^\perp$ to obtain the injection $v : N^\perp \to H$, then the bivariate function

$$\langle f, g \rangle_H := \langle v^{-1}(f), v^{-1}(g) \rangle_F \tag{3.5}$$

for $f, g \in H$ forms an inner product. Note that $N^\perp$ is a closed subspace of the Hilbert space F.

Proposition 37 *If the direct sum H of Hilbert spaces H_1, H_2 has the inner product (3.5), then H is complete (a Hilbert space).*

Proof. Since F is a Hilbert space (Proposition 36) and $N^\perp$ is its closed subset, $N^\perp$ is complete. Thus, we have

$$\|f_n - f_m\|_H \to 0 \Rightarrow \|v^{-1}(f_n - f_m)\|_F \to 0$$
$$\Rightarrow g \in F \text{ exists such that } \|v^{-1}(f_n) - g\|_F \to 0$$
$$\Rightarrow \|f_n - v(g)\|_H \to 0, v(g) \in H .$$

                                                                                     □

Proposition 38 (Aronszajn[1]) *Let k_1, k_2 be the reproducing kernels of RKHSs H_1, H_2, respectively. Then, $k = k_1 + k_2$ is the reproducing kernel of the Hilbert space*

$$H := H_1 \oplus H_2 := \{f_1 + f_2 | f_1 \in H_1, f_2 \in H_2\}$$

such that the inner product is (3.5) and the norm is

$$\|f\|_H^2 = \min_{f=f_1+f_2, f_1 \in H_1, f_2 \in H_2} \{\|f_1\|_{H_1}^2 + \|f_2\|_{H_2}^2\} \tag{3.6}$$

for $f \in H$.

Proof. The proof proceeds as follows:

1. Let $f \in H$ and $N^\perp \ni (f_1, f_2) := v^{-1}(f)$. We define $k(x, \cdot) := k_1(x, \cdot) + k_2(x, \cdot)$ and $(h_1(x, \cdot), h_2(x, \cdot)) := v^{-1}(k(x, \cdot))$, and we show that

$$\langle f_1, h_1(x, \cdot)\rangle_1 + \langle f_2, h_2(x, \cdot)\rangle_2 = \langle f_1, k_1(x, \cdot)\rangle_1 + \langle f_2, k_2(x, \cdot)\rangle_2 \,.$$

2. Using the above, we present the reproducing property $\langle f, k(x, \cdot)\rangle_H = f(x)$ of k.
3. We show that the norm of H is (3.6).

For details, see the appendix at the end of this chapter. $\qquad\square$

In the following, we construct the Sobolev space as an example of an RKHS and obtain its kernel.

Let $W_1[0, 1]$ be the set of f's defined over $[0, 1]$ such that f is differentiable almost everywhere and $f' \in L^2[0, 1]$. Then, we can write each $f \in W_1[0, 1]$ as

$$f(x) = f(0) + \int_0^x f'(y)dy \,. \tag{3.7}$$

Similarly, let $W_q[0, 1]$ be the set of f's defined over $[0, 1]$ such that f is differentiable $q - 1$ times and q times almost everywhere and $f^{(q)} \in L^2[0, 1]$. If we define

$$\phi_i(x) := \frac{x^i}{i!} \,, \quad i = 0, 1, \ldots$$

and

$$G_q(x, y) := \frac{(x - y)_+^{q-1}}{(q - 1)!} \,,$$

then we can Taylor-expand each $f \in W_q[0, 1]$ as follows:

$$f(x) = \sum_{i=0}^{q-1} f^{(i)}(0)\phi_i(x) + \int_0^1 G_q(x, y)f^{(q)}(y)dy. \tag{3.8}$$

In fact, we have the partial integral

$$\int_0^1 G_q(x, y) f^{(q)}(y) dy = \left[G_q(x, y) f^{(q-1)}(y) \right]_0^1 - \int_0^1 \{ \frac{d}{dy} G_q(x, y) \} f^{(q-1)}(y) dy$$

$$= -\frac{x^{q-1}}{(q-1)!} f^{(q-1)}(0) + \int_0^1 G_{q-1}(x, y) f^{(q-1)}(y) dy$$

and obtain (3.7) by repeatedly applying this integral to the right-hand side of (3.8).
For the transformation, we use

$$\int_0^1 G_q(x, y) h(y) dy = \int_0^1 \frac{(x-y)_+^{q-1}}{(q-1)!} h(y) dy$$

$$= \frac{1}{(q-1)!} \sum_{i=0}^{q-1} \binom{q-1}{i} x^i \int_0^x (-y)^{q-1-i} h(y) dy$$

and the differentiation

$$\int_0^1 \frac{d}{dy} \{ G_q(x, y) h(y) \} dy = \frac{1}{(q-2)!} \sum_{i=0}^{q-2} \binom{q-2}{i} \{ -x^i \int_0^x (-y)^{q-2-i} h(y) dy \}$$

$$= -\int_0^1 G_{q-1}(x, y) h(y) dy \ .$$

Hereafter, we write each element of $W_q[0, 1]$ as

$$\sum_{i=0}^{q-1} \alpha_i \phi_i(x) + \int_0^1 G_q(x, y) h(y) dy \tag{3.9}$$

$\alpha_0 = f(0), \ldots, \alpha_{q-1} = f^{(q-1)}(0) \in \mathbb{R}, h \in L^2[0, 1]$.

Although more than one Hilbert space $W_q[0, 1]$ exists with different definitions of inner products, we consider the Hilbert space H that can be written as the direct sum of H_0 and H_1, which is defined below. Let

$$H_0 := \text{span}\{\phi_0, \ldots, \phi_{q-1}\},$$

and define its inner product by

$$\langle f, g \rangle_{H_0} = \sum_{i=0}^{q-1} f^{(i)}(0) g^{(i)}(0)$$

for $f, g \in H_0$. We find that the inner product $\langle \cdot, \cdot \rangle_{H_0}$ satisfies the requirement of inner products and that $\{\phi_0, \ldots, \phi_{q-1}\}$ is an orthonormal basis. Since the inner product space H_0 is of finite dimensionality, it is apparently a Hilbert space. We define another inner product space H_1 as

$$H_1 := \{ \int_0^1 G_q(x, y)h(y)dy \mid h \in L^2[0, 1] \} \ .$$

Since $h \in L^2[0, 1]$, if we define the inner product as

$$\langle f, g \rangle_{H_1} = \int_0^1 f^{(q)}(y)g^{(q)}(y)dy$$

for $f, g \in H$, then we have

$$\| f_m - f_n \|_{H_1} \to 0 \iff \| f_m^{(q)} - f_n^{(q)} \|_{L^2[0,1]} \to 0,$$

and there exists an $f \in H_1$ such that

$$\| f_n - f \|_{H_1} \to 0 \iff \| f_n^{(q)} - f^{(q)} \|_{L^2[0,1]} \to 0 \ .$$

From Proposition 14, we have $\| f_m - f_n \|_{H_1} \to 0 \Rightarrow \| f_n - f \|_{H_1} \to 0$ (completeness), and H_1 is a Hilbert space. Moreover, from

$$f(x) = \sum_{i=0}^{q-1} \alpha_i \phi_i(x) \in H_1 \Rightarrow h = f^{(q)} = 0$$

and

$$f(x) = \int_0^1 G_q(x, y)h(y)dy \in H_0 \Rightarrow \alpha_0 = f(0) = 0, \dots, \alpha_{q-1} = f^{(q-1)}(0) = 0 \ ,$$

we have that $H_0 \cap H_1 = \{0\}$. From Proposition 38, for $f = f_0 + f_1$, $g = g_0 + g_1$, $f_0, g_0 \in H_0$, and $f_1, g_1 \in H_1$, the inner product is

$$\langle f, g \rangle_{W_q[0,1]} = \langle f_0 + f_1, g_0 + g_1 \rangle_{W_q[0,1]} = \langle f_0, g_0 \rangle_{H_0} + \langle f_1, g_1 \rangle_{H_1} \ .$$

The reproducing kernels of H_0, H_1 are, respectively,

$$k_0(x, y) := \sum_{i=0}^{q-1} \phi_i(x)\phi_i(y)$$

and

$$k_1(x, y) := \int_0^1 G_q(x, z)G_q(y, z)dz \ ,$$

where k_0 is derived from Example 3.2, and k_1 is derived from

$$\langle f(\cdot), k_1(x, \cdot)\rangle_{H_1} = \langle \int_0^1 G_q(\cdot, z)h(z)dz, \int_0^1 G_q(x, z)G_q(\cdot, z)dz\rangle_{H_1}$$
$$= \int_0^1 G_q(x, z)h(z)dz = f(x)$$

for arbitrary $f(\cdot) = \int_0^1 G_q(\cdot, z)h(z)dz \in H$ and $x \in E$ (the uniqueness is due to Proposition 32).

Furthermore, we can construct $W_q[0, 1]$ such that its kernel is

$$k(x, y) = k_0(x, y) + k_1(x, y)$$

for $x, y \in E$.

3.3 Mercer's Theorem

Let $(E, \mathcal{F}, \mu)$ be a measure space. We assume that the integral operator kernel $K : E \times E \to \mathbb{R}$ is a measurable function and is not necessarily nonnegative definite.

Suppose that $\int\int_{E \times E} K^2(x, y)d\mu(x)d\mu(y)$ takes finite values. Then, we define the integral operator T_K by

$$(T_K f)(\cdot) := \int_E K(x, \cdot)f(x)d\mu(x) \tag{3.10}$$

for $f \in L^2(E, \mathcal{B}, \mu)$. Since

$$\|T_K f\|^2 = \int_E \{(T_K f)(x)\}^2 d\mu(x) \le \int\int_{E \times E} \{K(x, y)\}^2 d\mu(x)d\mu(y) \int_E \{f(z)\}^2 d\mu(z)$$
$$= \|f\|^2 \int\int_{E \times E} \{K(x, y)\}^2 d\mu(x)d\mu(y),$$

we have $T_K \in B(L^2(E, \mathcal{B}, \mu))$ and

$$\|T_K\| \le \left(\int\int_{E \times E} K^2(x, y)d\mu(x)d\mu(y) \right)^{1/2}.$$

In the following, we assume that $K : E \times E \to \mathbb{R}$ is continuous and that the entire set E is compact (such as $E = [0, 1]$). Thus, we assume that the integral operator kernel K is uniformly continuous (Proposition 8).

Lemma 1 *For each $f \in L^2(E, \mathcal{F}, \mu)$, $T_K f(\cdot)$ is uniformly continuous.*

Proof. Since $E \times E \to \mathbb{R}$ is uniformly continuous, we achieve $|K(x, y) - K(x, z)| < \epsilon$ by making $|y - z|$ smaller for arbitrary $x \in E$ and $\epsilon > 0$. Thus, we have

$$\left| \int_E K(x, y) f(x) d\mu(x) - \int_E K(x, z) f(x) d\mu(x) \right| \leq \epsilon \|f\| .$$

$\square$

Proposition 39 *T_K is a compact operator.*

Proof. By Proposition 12, for an arbitrary $\epsilon > 0$, there exist $n(\epsilon) \geq 1$ and an $\mathbb{R}$-coefficient bivariate polynomial $K_{n(\epsilon)}(x, y) := \sum_{i=1}^{n(\epsilon)} g_i(x) y^i$ whose order of y is at most $n(\epsilon)$ such that

$$\sup_{x, y \in E} |K(x, y) - K_{n(\epsilon)}(x, y)| < \epsilon ,$$

where $g_1, \ldots, g_{n(\epsilon)}$ are $\mathbb{R}$-coefficient univariable polynomials. If we abbreviate $n(\epsilon)$ as n and write the integral operator corresponding to K_n as T_{K_n}, then we may regard

$$T_{K_n} f(\cdot) = \sum_{i=0}^{n} y^i \int_E f(x) g_i(x) d\mu(x)$$

as

$$T_{K_n} f : H \ni f \mapsto \left[\int_E f(x) g_0(x) d\mu(x), \ldots, \int_E f(x) g_n(x) d\mu(x) \right] \in \mathbb{R}^{n+1} .$$

Since the rank of T_{K_n} is finite, from the first item of Proposition 25, T_{K_n} is a compact operator. Moreover, since

$$\|(T_{K_n} - T_K) f\|^2 = \int_E \left(\int_E [K_n(x, y) - K(x, y)] f(y) d\mu(y) \right)^2 d\mu(x) \leq \epsilon^2 \|f\|^2 \mu^2(E) ,$$

from Proposition 25, T_K is a compact operator. $\square$

In the following, we assume that K is symmetric. Then, from Example 45, T_K is self-adjoint. Thus, from Proposition 39, we have that

$$T_K x = \sum_{j=1}^{\infty} \lambda_j \langle e_j, x \rangle e_j$$

using $\{\lambda_j\}$ and $\{e_j\}$ that satisfy Proposition 27. Moreover, Lemma 1 implies the following:

Lemma 2

$$e_j(y) = \lambda_j^{-1} \int_E K(x, y) e_j(x) d\mu(x)$$

is uniformly continuous w.r.t. y.

Example 58 (*Brown Motion*) We obtain the eigenvalues and eigenfunctions $\{(\lambda_j, e_j)\}$ when the integral operator kernel in $L^2[0, 1]$ is $K(x, y) = \min\{x, y\}$, $x, y \in E = [0, 1]$, (the subspace H_1 of the Sobolev space $W_1[0, 1]$). Since

$$T_K f(x) = \int_0^1 K(x, y) f(y) dy = \int_0^x y f(y) dy + x \int_x^1 f(y) dy ,$$

the eigenequation is

$$\int_0^1 \min(x, y) e(y) dy = \lambda e(x) , \qquad (3.11)$$

i.e.,

$$\int_0^x y e(y) dy + x \int_x^1 e(y) dy = \lambda e(x) .$$

If we differentiate the both sides by x, we obtain

$$x e(x) + \int_x^1 e(y) dy - x e(x) = \lambda e'(x) ,$$

i.e.,

$$\int_x^1 e(y) dy = \lambda e'(x) . \qquad (3.12)$$

If we further differentiate both sides by x, then we obtain $e(x) = -\lambda e''(x)$ and

$$e(y) = \alpha \sin(y/\sqrt{\lambda}) + \beta \cos(y/\sqrt{\lambda}) .$$

If we substitute $x = 0$ into (3.11), then we have $e(0) = 0$, which is equivalent to $\beta = 0$. From (3.12), we have $e'(1) = 0$, i.e., $\alpha \cos(1/\sqrt{\lambda}) = 0$. Thus, we obtain

$$1/\sqrt{\lambda} = (2j - 1)\pi/2 , \quad j = 1, 2, \ldots .$$

Therefore, the eigenvalues are

$$\lambda_j = \frac{4}{\{(2j - 1)\pi\}^2} , \qquad (3.13)$$

and the orthonormal eigenfunctions are

$$e_j(x) = \sqrt{2} \sin\left(\frac{(2j-1)\pi}{2}x\right), \tag{3.14}$$

where to derive $\alpha = \sqrt{2}$, we use

$$\int_0^1 \sin^2\left(\frac{y}{\sqrt{\lambda}}\right) dy = \int_0^1 \frac{1 - \cos\left(\frac{2y}{\sqrt{\lambda}}\right)}{2} dy = \frac{1}{2} - \frac{1}{2}\left[\frac{\sqrt{\lambda}}{2}\sin\frac{2y}{\sqrt{\lambda}}\right]_0^1 = \frac{1}{2}.$$

Example 59 (*Zhu et al.* [36]) For a Gaussian kernel,

$$K(x, y) = \exp\left(\frac{-(x-y)^2}{2\sigma^2}\right)$$

if we regard the finite measure μ in (3.10) of the integral operator kernel as a Gaussian distribution with a mean of 0 and a variance of $\hat{\sigma}^2$, then the eigenvalue and eigenfunction are

$$\lambda_j = \sqrt{\frac{2a}{A}} B^j$$

and

$$e_j(x) = \exp(-(c-a)x^2)H_j(\sqrt{2c}x),$$

where H_j is a Hermite polynomial of order j:

$$H_j(x) := (-1)^j \exp(x^2)\frac{d^j}{dx^j}\exp(-x^2),$$

$a^{-1} := 4\hat{\sigma}^2$, $b^{-1} := 2\sigma^2$, $c := \sqrt{a^2 + 2ab}$, $A := a + b + c$, and $B := b/A$. The proof is not difficult but rather monotonous and long. See the appendix at the end of this chapter for details. Note that for a Gaussian kernel with a parameter σ^2, if the measure is also a Gaussian distribution with a mean of 0 and a variance of $\hat{\sigma}^2$, we can compute the eigenvalues from $\beta := \dfrac{\hat{\sigma}^2}{\sigma^2} = \dfrac{b}{2a}$:

$$\sqrt{\frac{2a}{A}} B^j = \sqrt{\frac{2a}{a + b + \sqrt{a^2 + 2ab}}}\left(\frac{b}{a + b + \sqrt{a^2 + 2ab}}\right)^j$$

$$= [1/2 + \beta + \sqrt{1/4 + \beta}]^{-1/2}\left(\frac{\beta}{1/2 + \beta + \sqrt{1/4 + \beta}}\right)^j,$$

which forms a geometric sequence. For example, if $\sigma^2 = \hat{\sigma}^2 = 1$, then the eigenvalue is

$$\lambda_j = \left(\frac{3 - \sqrt{5}}{2}\right)^{j+1/2}.$$

Fig. 3.1 The eigenfunctions for the Gaussian kernel and Gaussian distribution, where $\sigma^2 = \hat{\sigma}^2 = 1$. If j is odd, the eigenfunctions are even and odd functions, respectively

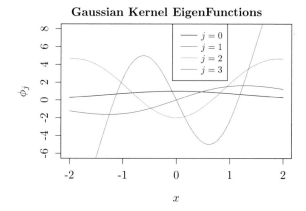

The Hermite polynomials are $H_1(x) = 2x$, $H_2(x) = -2 + 4x^2$, and $H_3(x) = 12x - 8x^3$ ($H_0(1) = 1$, $H_j(x) = 2x H_{j-1}(x) - H'_{j-1}(x)$), and the other quantities are

$$c = \sqrt{a^2 + 2ab} = (4\hat{\sigma}^2)^{-1}\sqrt{1 + 4\hat{\sigma}^2/\sigma^2} = \frac{\sqrt{5}}{4}, \quad a = (4\hat{\sigma}^2)^{-1} = \frac{1}{4}.$$

We show the eigenfunction ϕ_j for $j = 1, 2, 3$ in Fig. 3.1. The code is as follows:

```
Hermite=function(j){    ## The index starts from 1 in the R language
  if(j==0)return(1)
  a=rep(0,j+2); b=rep(0,j+2)
  a[1]=1
  for(i in 1:j){
    b[1]=-a[2]
    for(k in 1:(i+1))b[k+1]=2*a[k]-(k+1)*a[k+2]
    a=b
  }
  return(b[1:(j+1)])    ## Output the Coefficients of the Hermite polynomial
}
```

```
Hermite(2)              ## Hermite Polynomial of order 2
Hermite(3)              ## Hermite Polynomial of order 3
Hermite(4)              ## Hermite Polynomial of order 4
```

```
H=function(j,x){
  coef=Hermite(j)
  S=0
  for(i in 0:j) S=S+coef[i+1]*x^i
  return(S)
}
```

```
1  cc=sqrt(5)/4; a=1/4
2  phi=function(j,x) exp(-(cc-a)*x^2)*H(j,sqrt(2*cc)*x)
3  curve(phi(0,x),-2,2, ylim=c(-2,8),col=1,ylab="phi")
4  for(i in 1:3)curve(phi(i,x),-2,2, ylim=c(-2,8), add=TRUE, ann=FALSE,
       col=i+1)
5  legend("topright",legend=paste("j=",0:3),lwd=1, col=1:4)
6  title("The eigenfunction of Gaussian kernel")
```

In this section, we prove Mercer's theorem for integral operators and illustrate some examples. Hereafter, we assume that K and T_K are nonnegative definite.

Proposition 40 *An integral operator T_K is nonnegative definite if and only if $K : E \times E \to \mathbb{R}$ is nonnegative definite, i.e., K is a positive definite kernel.*

Proof: See the appendix at the end of this chapter.

Proposition 41 (Mercer [21]) *Let $K : E \times E \to \mathbb{R}$ be a continuous positive definite kernel and T_K be the corresponding integral operator. Let $\{(\lambda_j, e_j)\}_{j=1}^{\infty}$ be the sequence of eigenvalues and eigenvectors of T_K. Then, we can write*

$$K(x, y) = \sum_{j=1}^{\infty} \lambda_j e_j(x) e_j(y),$$

and this sum absolutely and uniformly converges.

By absolute convergence, we mean that the sum of the absolute values converges, and by uniform convergence, we mean that the upper bound of the error that does not depend on $x, y \in E$ converges to zero.

   Proof: Note that $K_n(x, y) := K(x, y) - \sum_{j=1}^{n} \lambda_j e_j(x) e_j(y)$ is continuous and that the integral operator T_{K_n} is nonnegative definite. In fact, for each $f \in L^2$ $(E, \mathcal{F}, \mu)$, we have

$$\langle T_{K_n} f, f \rangle = \langle T_K f, f \rangle - \sum_{j=1}^{n} \lambda_j \langle f, e_j \rangle^2 = \sum_{j=n+1}^{\infty} \lambda_j \langle f, e_j \rangle^2 \geq 0 .$$

Thus, from Proposition 40, K_n is nonnegative definite, and $K_n(x, x) \geq 0$. Thus, for all $x \in E$, we have

$$\sum_{j=1}^{\infty} \lambda_j e_j^2(x) \leq K(x, x) . \tag{3.15}$$

Moreover, for any set J consisting of positive numbers, we have

$$\sum_{j \in J} |\lambda_j e_j(x) e_j(y)| \leq \left(\sum_{j \in J} \lambda_j e_j^2(x) \right)^{1/2} \left(\sum_{j \in J} \lambda_j e_j^2(y) \right)^{1/2} , \tag{3.16}$$

which means that from (3.15),

$$\sum_{j\in J} |\lambda_j e_j(x) e_j(y)| \le \{K(x,x)K(y,y)\}^{1/2}$$

for $x, y \in E$. From (3.16), we have

$$\sum_{j=n+1}^{\infty} |\lambda_j e_j(x) e_j(y)| \le \left(\sum_{j=n+1}^{\infty} \lambda_j e_j^2(x) \right)^{1/2} \left(\sum_{j=n+1}^{\infty} \lambda_j e_j^2(y) \right)^{1/2}$$

and the right-hand side monotonically converges to 0 as n grows. Since E is compact, the left-hand side uniformly converges according to the lemma below.

Lemma 3 (Dini) *Let E be a compact set. For a continuous function $f_n : E \to \mathbb{R}$, if $f_n(x)$ monotonically converges to $f(x)$ for a continuous f and each $x \in E$, then the convergence is uniform.*

Proof. See the appendix at the end of this chapter.
   Thus, for an arbitrary $\epsilon > 0$, there exists an n such that

$$\sup_{x,y\in E} \sum_{j=n+1}^{\infty} |\lambda_j e_j(x) e_j(y)| < \epsilon, \tag{3.17}$$

and this sum absolutely and uniformly converges.   □

Example 60 (*The Kernel Expressed by the Difference Between Two Variables*) Let $E = [-1, 1]$. An integral operator for which $K : E \times E \to \mathbb{R}$ can be expressed by $K(x,z) = \phi(x-z)$ ($\phi : E \to \mathbb{R}$) is $T_K f(x) = \int_E \phi(x-y)f(y)dy$, which can be expressed by $(\phi * f)(x)$ using convolution: $(g * h)(u) = \int_E g(u-v)h(v)$. Hereafter, we assume that the cycle of ϕ is two, i.e., $\phi(x) = \phi(x + 2\mathbb{Z})$. In this case, $e_j(x) = \cos(\pi j x)$ is the eigenfunction of T_K. In fact, since ϕ is an even function and is cyclic, we have

$$T_K e_j(x) = \int_E \phi(x-y)\cos(\pi j y)dy = \int_{-1-x}^{1-x} \phi(-u)\cos(\pi j(x+u))du = \int_E \phi(u)\cos(\pi j(x+u))du$$

and

$$T_K e_j(x) = \{\int_E \phi(u)\cos(\pi j u)du\}\cos(\pi j x) - \{\int_E \phi(u)\sin(\pi j u)du\}\sin(\pi j x)$$
$$= \lambda_j \cos(\pi j x)$$

from the Addition theorem $\cos(\pi j(x+u)) = \cos(\pi j x)\cos(\pi j u) - \sin(\pi j x)$ $\sin(\pi j u)$, where $\lambda_j = \int_E \phi(u)\cos(\pi j u)du$. Similarly, $\sin(\pi j x)$ is an eigenfunction, and λ_j is the corresponding eigenvalue. Thus, from Mercer's theorem, we have

$$K(x, y) = \sum_{j=0}^{\infty} \lambda_j \{\cos(\pi j x)\cos(\pi j y) + \sin(\pi j x)\sin(\pi j y)\} = \sum_{j=0}^{\infty} \lambda_j \cos\{\pi j (x - y)\} \,.$$

Example 61 (*Polynomial Kernel*) For the polynomial kernel in Example 8, let $m = 2, d = 1$. We compute the eigenfunction of $K(x, y) = (1 + xy)^2$ over $x, y \in E = [-1, 1]$ by setting $e(x) := a_0 + a_1 x + a_2 x^2$. By comparing

$$\int_E K(x, y)e(y)dy = \int_E (1 + xy)^2 e(y)dy = \int_E e(y)dy + \{2\int_E ye(y)dy\}x + \{\int_E y^2 f(y)dy\}x^2$$

with $\lambda e(x)$, we obtain

$$\begin{cases} \int_E (a_0 + a_1 y + a_2 y^2)dy & = \lambda a_0 \\ 2\int_E y(a_0 + a_1 y + a_2 y^2)dy = \lambda a_1 \\ \int_E y^2(a_0 + a_1 y + a_2 y^2)dy & = \lambda a_2 \end{cases} .$$

We solve the eigenequation w.r.t. the following matrix:

$$\begin{bmatrix} \int_E dy & \int_E ydy & \int_E y^2 dy \\ 2\int_E ydy & 2\int_E y^2 dy & 2\int_E y^3 dy \\ \int_E y^2 dy & \int_E y^3 dy & \int_E y^4 dy \end{bmatrix} \begin{bmatrix} a_0 \\ a_1 \\ a_2 \end{bmatrix} = \lambda \begin{bmatrix} a_0 \\ a_1 \\ a_2 \end{bmatrix} .$$

Now, we consider the general method for approximately obtaining eigenvalues and eigenvectors in Mercer's theorem. Let X be a random variable in E. Then, for the integral operator $T_x \in B(H)$ $(x \in E)$ defined by

$$T_K : L^2 \ni \phi \mapsto \int_E K(\cdot, x)\phi(x)d\mu(x) \in L^2 \,,$$

there exist $\lambda_1 \geq \lambda_2 \geq \ldots$ and $\phi_1, \phi_2, \ldots \in L^2$ such that

$$T_K \phi_j = \lambda \phi_j$$

and

$$\int_E \phi_j \phi_k d\mu = \delta_{j,k} \,.$$

We say that the probability μ has generated $x_1, \ldots, x_m \in E$ with $m \geq 1$, and we approximate the generation as

$$\frac{1}{m}\sum_{j=1}^{m} K(x_j, y)\phi_i(x_j) = \lambda_i \phi_i(y) \,, \quad y \in E \tag{3.18}$$

$i = 1, 2, \ldots$. Since we have

$$\frac{1}{m} \sum_{i=1}^{m} \phi_j(x_i)\phi_k(x_i) = \delta_{j,k}$$

if we substitute $x_1, \ldots, x_m$ into y in (3.18), we find that there exists an orthogonal matrix $U \in \mathbb{R}^{m \times m}$ such that

$$K_m U = U \Lambda ,$$

where $K_m \in \mathbb{R}^{m \times m}$ is the Gram matrix and Λ is the diagonal matrix with the elements $\lambda_1^{(m)} = m\lambda_1, \ldots, \lambda_m^{(m)} = m\lambda_m$. If we substitute $\phi_i(x_j) = \sqrt{m}U_{j,i}$, $\lambda_i = \dfrac{\lambda_i^{(m)}}{m}$ into (3.18), we obtain

$$\phi_i(\cdot) = \frac{\sqrt{m}}{\lambda_i^{(m)}} \sum_{j=1}^{m} K(x_j, \cdot)U_{j,i} . \tag{3.19}$$

We require that the distribution of $x_1, \ldots, x_m \in E$ coincide with the measure μ of the integral operator. It is known that if we make m larger in $\lambda_i^{(m)}/m$, the term converges to the eigenvalue λ_i. For the proof and the convergence process, consult Baker (Theorem 3.4 [3]).

We write the procedure using the R language as below.

Example 62 We obtain the eigenvalue and eigenfunction by using the following program with a Gaussian kernel, where the measure required for the definition of the integral kernel should be the same as the measure used when providing random numbers. Even with the same Gaussian kernel, if $x_1, \ldots, x_N$ follows a different distribution, we obtain different eigenvalues and eigenfunctions. We compare the cases in which $N = 300$ and $N = 1000$ to find that the eigenvalues and eigenfunctions coincide (Figs. 3.2 and 3.3).

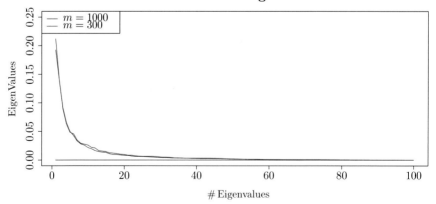

Fig. 3.2 The eigenvalues obtained in Example 62. We compare the cases involving $m = 1000$ samples and the first $m = 300$ samples. The largest eigenvalues for both cases coincide

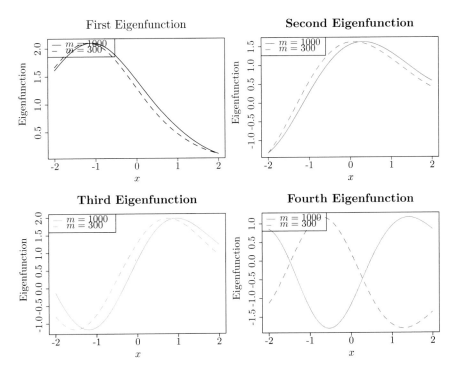

Fig. 3.3 The eigenfunctions obtained in Example 62. We show a comparison between the functions of the $m = 1000$ samples and the first $m = 300$ samples. The eigenfunctions coincide for the first largest three eigenvalues, but they are far from each other for the fourth eigenvalue. However, the fourth eigenvalues coincide

```
## Kernel definition
sigma=1; k=function(x,y)exp(-(x-y)^2/sigma^2)

## Generate m samples and its Gram matrix
m=300; x=rnorm(m)-2*rnorm(m)^2+3*rnorm(m)^3
K=matrix(0,m,m)
for(i in 1:m)for(j in 1:m)K[i,j]=k(x[i],x[j])

## Eigenvalues and Eigenvectors
eig=eigen(K)
lam.m=eig$values
lam=lam.m/m
U=eig$vector
alpha=array(0,dim=c(m,m))
for(i in 1:m)alpha[,i]=U[,i]*sqrt(m)/lam.m[i]

## Display Graph
F=function(y,i){
  S=0; for(j in 1:m)S=S+alpha[j,i]*k(x[j],y)
  return(S)
}
```

```
22  i=1    ## Change i and execute it
23  G=function(y)F(y,i)
24  plot(G,xlim=c(-2,2))
25  title("Eigen Values and their Eigen Functions")
```

Finally, we present the RKHS obtained from Mercer's theorem (Proposition 41). In Example 57, we pointed out that the condition was too loose for the L^2-space to be an RKHS. The following proposition suggests the restrictions that we should add.

Proposition 42 *Let $\{(\lambda_j, e_j)\}$ be an eigenvalue of an integral operator with a positive definite kernel k and an orthonormal eigenfunction. In this case,*

$$H = \{\sum_{j=1}^{\infty} \beta_j e_j | \sum_{j=1}^{\infty} \frac{\beta_j^2}{\lambda_j} < \infty\}$$

$$\langle f, g \rangle_H := \sum_{j=1}^{\infty} \frac{\int_E f(x)e_j(x)d\eta(x) \int_E g(x)e_j(x)d\eta(x)}{\lambda_j} \quad (3.20)$$

gives the RKHS.

The proposition claims that if we restrict the elements $\sum_{j=1}^{\infty} \beta_j e_j$ for which $\sum_{j=1}^{\infty} \beta_j^2 < \infty$ to those for which $\sum_{j=1}^{\infty} \frac{\beta_j^2}{\lambda_j} < \infty$, the L^2 space becomes an RKHS.

Proof. From the definition of the inner product (3.20), we can write $\langle e_i, e_j \rangle_H = \frac{1}{\lambda_i}\delta_{i,j}$. Thus, we have

$$\int_E \{\sum_{j=1}^{\infty} \beta_j e_j(x)\}^2 d\beta(x) < \infty \iff \sum_{j=1}^{\infty} \frac{\beta_j^2}{\lambda_j} < \infty ,$$

and H is a Hilbert space. From Mercer's theorem, we can write $k(x, \cdot) = \sum_{j=1}^{\infty} \lambda_j e_j(x)e_j(\cdot)$, so we have

$$\sum_{j=1}^{\infty} \frac{\{\lambda_j e_j(x)\}^2}{\lambda_j} = \sum_{j=1}^{\infty} \lambda_j e_j(x)e_j(x) = k(x, x) < \infty$$

and $k(x, \cdot) \in H$. Finally, since $\int_E k(\cdot, y)e_j(y)d\eta(y) = \lambda_j e_j(\cdot)$, we have

$$\langle f, k(\cdot, x) \rangle_H = \sum_{j=1}^{\infty} \frac{1}{\lambda_j} \int_E f(y)e_j(y)d\eta(y) \int_E k(x, y)e_j(y)d\eta(y)$$

$$= \sum_{j=1}^{\infty} \{\int_E f(y)e_j(y)d\eta(y)\}e_j(x) = f(x) ,$$

which is the reproducing property.   □

As seen from the proof, the eigenvector $\{e_j\}$ of Mercer's theorem is orthonormal in the L^2 space, but in the obtained RKHS, the norm is $\lambda_j^{-1/2}$. We can see that the RKHS reduces $\{\beta_j\}$ faster than the L^2 space.

Appendix

Proof of Proposition 34

Let $k : E \times E \to \mathbb{R}$ be the positive definite kernel of a Hilbert space H. We show that for the linear space H_0 spanned by $k(x, \cdot)$, $x \in E$, the bivariate function

$$\langle f, g \rangle_{H_0} = \sum_{i=1}^{m} \sum_{j=1}^{n} a_i b_j k(x_i, y_j)$$

is an inner product between

$$f(\cdot) = \sum_{i=1}^{m} a_i k(x_i, \cdot) \text{ and } g(\cdot) = \sum_{j=1}^{n} b_j k(y_j, \cdot) \in H_0 . \tag{3.21}$$

$$\langle f, g \rangle_{H_0} = \sum_{i=1}^{m} a_i g(x_i) = \sum_{j=1}^{m} b_j f(x_j)$$

does not depend on the choice of f, g in (3.21). In particular, $\langle f, g \rangle_{H_0}$ is symmetric. Since k is a positive definite kernel, we have

$$\|f\|^2 = \sum_{i=1}^{m} \sum_{j=1}^{n} a_i a_j k(x_i, x_j) \geq 0 .$$

Moreover, from

$$|f(x)| = |\langle f(\cdot), k(x, \cdot) \rangle_{H_0}| \leq \|f\|_{H_0} \sqrt{k(x, x)} ,$$

we have $\|f\|_{H_0} = 0 \Rightarrow f = 0$. In the following, we construct the linear space H obtained by completing H_0.

Let $\{f_n\}$ be a Cauchy sequence in H_0. For an arbitrary $x \in E$ and $m, n \geq 1$, we have

$$|f_m(x) - f_n(x)| \leq \|f_m - f_n\|_{H_0} \sqrt{k(x, x)} ,$$

and $\{f_n(x)\}$ is Cauchy. Since this sequence is a real sequence, it has a convergence point for each $x \in E$. In the following, let H be the set of $f : E \to \mathbb{R}$ such that the $\{f_n(x)\}$ for which $\{f_n\}$ is Cauchy in H_0 converges to $f(x)$ for each $x \in E$. In general,

H_0 is a subset of H. In the following, we define an inner product in H and prove that H is an RKHS with a reproducing kernel k.

Lemma 4 *Suppose that $\{f_n\}$ is a Cauchy sequence in H_0. If the sequence $\{f_n(x)\}$ converges to 0 for each $x \in E$, then we have*

$$\lim_{n \to \infty} \|f_n\|_{H_0} = 0 .$$

Proof of Lemma 4: Since a Cauchy sequence is bounded (Example 26), there exists a $B > 0$ such that $\|f_n\| < B$, $n = 1, 2, \ldots$. Moreover, since the above sequence is a Cauchy sequence, for an arbitrary $\epsilon > 0$, there exists an N such that $n > N \Rightarrow \|f_n - f_N\| < \epsilon/B$. Thus, for $f_N(x) = \sum_{i=1}^{p} \alpha_i k(x_i, x) \in H_0$, $\alpha_i \in \mathbb{R}$, $x_i \in E$, and $i = 1, 2, \ldots$, we have that when $n > N$

$$\|f_n\|_{H_0}^2 = \langle f_n - f_N, f_n \rangle_{H_0} + \langle f_N, f_n \rangle_{H_0} \le \|f_n - f_N\|_{H_0}\|f_n\|_{H_0} + \sum_{i=1}^{p} \alpha_i |f_n(x_i)| .$$

Each of the first and second terms is at most ϵ since we have $f_n(x_i) \to 0$ as $n \to \infty$ for each $i = 1, \ldots, p$. Hence, we have Lemma 4.   □

For Cauchy sequences $\{f_n\}, \{g_n\}$ in H_0, we define $f, g \in H$ such that $\{f_n(x)\}$, $\{g_n(x)\}$ converge to $f(x), g(x)$, respectively, for each $x \in E$. Then, $\{\langle f_n, g_n \rangle_{H_0}\}$ is Cauchy:

$$|\langle f_n, g_n \rangle_{H_0} - \langle f_m, g_m \rangle_{H_0}| = |\langle f_n, g_n - g_m \rangle_{H_0} + \langle f_n - f_m, g_m \rangle_{H_0}|$$
$$\le \|f_n\|_{H_0}\|g_n - g_m\|_{H_0} + \|f_n - f_m\|_{H_0}\|g_m\|_{H_0} .$$

Since $\{\langle f_n, g_n \rangle_{H_0}\}$ is real and Cauchy, it converges (Proposition 6). The inner product obtained by convergence depends only on $f(x), g(x)$ ($x \in E$).

Let $\{f_n'\}, \{g_n'\}$ be other Cauchy sequences in H_0 that converge to f, g for each $x \in E$. Then, $\{f_n - f_n'\}, \{g_n - g_n'\}$ are Cauchy sequences that converge to 0 for each $x \in E$, and from Lemma 4, we have $\|f_n - f_n'\|_{H_0}, \|g_n - g_n'\|_{H_0} \to 0$ as $n \to \infty$, which means that

$$|\langle f_n, g_n \rangle_{H_0} - \langle f_n', g_n' \rangle_{H_0}| = |\langle f_n, g_n - g_n' \rangle_{H_0} + \langle f_n - f_n', g_n' \rangle_{H_0}|$$
$$\le \|f_n\|_{H_0}\|g_n - g_n'\|_{H_0} + \|f_n - f_n'\|_{H_0}\|g_n'\|_{H_0} \to 0 .$$

Thus, the convergence point of $\{\langle f_n, g_n \rangle_{H_0}\}$ does not depend on $\{f_n\}, \{g_n\}$ but on $f, g \in H$. We define the inner product of H by

$$\langle f, g \rangle_H := \lim_{n \to \infty} \langle f_n, g_n \rangle_{H_0} .$$

To show that this expression satisfies the definition of an inner product, we assume that $\|f\|_H = \langle f, f \rangle_H = 0$. Then, for each $x \in E$, as $n \to \infty$, from

$$|f_n(x)| = |\langle f_n(\cdot), k(x, \cdot)\rangle| \leq \sqrt{k(x, x)}\|f_n\|_{H_0} \to 0 ,$$

we have $|f(x)| = \lim_{n\to\infty} |f_n(x)| = 0$.

Moreover, since we have defined $f \in H$ according to $\lim_{n\to\infty} f_n(x)$ $(x \in E)$ for any Cauchy sequence $\{f_n\}$ in H_0 that converges to f, from the definition of inner products, we have

$$\|f - f_n\|_H = \lim_{m\to\infty} \|f_m - f_n\|_{H_0} \to 0 \tag{3.22}$$

$n \to \infty$, and H_0 is dense in H.

We show that H is complete. Let $\{f_n\}$ be a Cauchy sequence in H. From denseness, there exists a sequence $\{f_n'\}$ in H_0 such that

$$\|f_n - f_n'\|_H \to 0 \tag{3.23}$$

as $n \to \infty$. Therefore, given an arbitrary $\epsilon > 0$, for $m, n > N$, we have $\|f_n - f_n'\|_H, \|f_m - f_m'\|_H, \|f_n - f_m\|_H < \epsilon/3$ and

$$\|f_n' - f_m'\|_{H_0} = \|f_n' - f_m'\|_H \leq \|f_n - f_n'\|_H + \|f_n - f_m\|_H + \|f_m - f_m'\|_H \leq \epsilon$$

for $f_n', f_m' \in H_0 \subseteq H$. Thus, $\{f_n'\}$ is a Cauchy sequence in H_0, and we define $f \in H$ by the convergence of $f(x)$ for each $x \in E$. Moreover, from (3.22), we have $\|f - f_n'\|_H \to 0$. Combining this with (3.23), we obtain

$$\|f - f_n\|_H \leq \|f - f_n'\|_H + \|f_n' - f_n\|_H \to 0$$

as $n \to \infty$. Hence, H is complete.

Next, we show that k is the corresponding reproducing kernel of the Hilbert space H. Property (3.1) holds immediately because $k(x, \cdot) \in H_0 \subseteq H, x \in E$. For another property (3.2), since $f \in H$ is a limit of the Cauchy sequence $\{f_n\}$ in H_0 at $x \in E$, we have

$$f(x) = \lim_{n\to\infty} f_n(x) = \lim_{n\to\infty} \langle f_n(\cdot), k(x, \cdot)\rangle_{H_0} = \langle f, k(x, \cdot)\rangle_H .$$

Finally, we show that such an H uniquely exists. Suppose that G exists and shares the same properties possessed by H. Since H is a closure of H_0, G should contain H as a subspace. Since H is closed, from (2.11), we write $G = H \oplus H^\perp$. However, since $k(x, \cdot) \in H, x \in E$ and $\langle f(\cdot), k(x, \cdot)\rangle_G = 0$ for $f \in H^\perp$, we have $f(x) = 0$, $x \in E$, which means that $H^\perp = \{0\}$.   □

Proof of Proposition 38

From our assumption, we have $k(x, \cdot) = k_1(x, \cdot) + k_2(x, \cdot) \in H$ for each $x \in E$. We define $\quad N^\perp \ni (h_1(x, \cdot), h_2(x, \cdot)) := v^{-1}(k(x, \cdot)) \quad$ for each $x \in E$, where $h_1(x, \cdot), h_2(x, \cdot)$ are elements in H_1, H_2 for $x \in E$, but h_1, h_2 are not necessarily reproducing kernels k_1, k_2 of H_1, H_2, respectively. Since $k(x, \cdot) = k_1(x, \cdot) +$

$k_2(x, \cdot)$, we have

$$h_1(x, \cdot) - k_1(x, \cdot) + h_2(x, \cdot) - k_2(x, \cdot) = k(x, \cdot) - k(x, \cdot) = 0$$

and $z := (h_1(x, \cdot) - k_1(x, \cdot), h_2(x, \cdot) - k_2(x, \cdot)) \in N$, so

$$0 = \langle 0, f \rangle_H = \langle z, (f_1, f_2) \rangle_F$$

for $f \in H$ and $N^\perp \ni (f_1, f_2) := v^{-1}(f)$. Thus, we have

$$\langle f_1, h_1(x, \cdot) \rangle_1 + \langle f_2, h_2(x, \cdot) \rangle_2 = \langle f_1, k_1(x, \cdot) \rangle_1 + \langle f_2, k_2(x, \cdot) \rangle_2,$$

which implies the reproducing property:

$$\langle f, k(x, \cdot) \rangle_H = \langle v^{-1}(f), v^{-1}(k(x, \cdot)) \rangle_F = \langle (f_1, f_2), (h_1(x, \cdot), h_2(x, \cdot)) \rangle_F$$
$$= \langle (f_1, f_2), (k_1(x, \cdot), k_2(x, \cdot)) \rangle_F = f_1(x) + f_2(x) = f(x).$$

Furthermore, let $(f_1, f_2) \in F$, $f := f_1 + f_2$, and $(g_1, g_2) := (f_1, f_2) - v^{-1}(f)$. Then, from $(g_1, g_2) \in N$ and $v^{-1}(f) \in N^\perp$, we have

$$\|(f_1, f_2)\|_F^2 = \|v^{-1}(f)\|_F^2 + \|(g_1, g_2)\|_F^2.$$

Combining this with (3.4) and (3.5), we have

$$\|f\|_H^2 = \|v^{-1}(f)\|_F^2 \leq \|(f_1, f_2)\|_F^2 = \|f_1\|_{H_1}^2 + \|f_2\|_{H_2}^2,$$

where the equality holds when $(f_1, f_2) = v^{-1}(f)$. $\square$

Proof of Example 59

We use the equality [10]

$$\int_{-\infty}^{\infty} \exp(-(x - y)^2) H_j(\alpha x) dx = \sqrt{\pi}(1 - \alpha^2)^{j/2} H_j\left(\frac{\alpha y}{(1 - \alpha^2)^{1/2}}\right).$$

Suppose that $\int_E p(y) dy = 1$. If we have

$$\int_E k(x, y) \phi_j(y) p(y) dy = \lambda \phi_j(x),$$

then

$$\int_E \tilde{k}(x, y) \tilde{\phi}_j(y) dy = \lambda \tilde{\phi}_j(x)$$

for $\tilde{k}(x, y) := p(x)^{1/2}k(x, y)p(y)^{1/2}$, $\tilde{\phi}_j(x) := p(x)^{1/2}\phi_j(x)$. Thus, it is sufficient to show that we obtain the right-hand side by substituting

$$p(x) := \sqrt{\frac{2a}{\pi}} \exp(-2ax^2)$$

$$\tilde{k}(x, y) := \sqrt{\frac{2a}{\pi}} \exp(-ax^2) \exp(-b(x - y)^2) \exp(-ay^2)$$

$$\tilde{\phi}_j(x) := (\frac{2a}{\pi})^{1/4} \exp(-cx^2) H_j(\sqrt{2c}x)$$

into the left-hand side for $E = (-\infty, \infty)$. The left-hand side becomes

$$\int_{-\infty}^{\infty} (\frac{2a}{\pi})^{3/4} \exp(-ax^2) \exp(-b(x - y)^2) \exp(-ay^2) \exp(-cy^2) H_j(\sqrt{2c}y) dy$$

$$= (\frac{2a}{\pi})^{3/4} \int_{-\infty}^{\infty} \exp\{-(a + b + c)(y - \frac{b}{a+b+c}x)^2 + [\frac{b^2}{a+b+c} - (a + b)]x^2\} H_j(\sqrt{2c}y) dy$$

$$= (\frac{2a}{\pi})^{3/4} \exp(-cx^2) \int_{-\infty}^{\infty} \exp\{-(z - \frac{b}{\sqrt{a+b+c}}x)^2\} H_j(\frac{\sqrt{2c}}{\sqrt{a+b+c}}z) \frac{dz}{\sqrt{a+b+c}}$$

$$= (\frac{2a}{\pi})^{1/4} \sqrt{\frac{2a}{\pi(a + b + c)}} \exp(-cx^2)\sqrt{\pi}(1 - \frac{2c}{a+b+c})^{j/2} H_j(\sqrt{2c}x)$$

$$= \sqrt{\frac{2a}{a + b + c}}(\frac{b}{a+b+c})^j (\frac{2a}{\pi})^{1/4} \exp(-cx^2) H_j(\sqrt{2c}x) = \sqrt{\frac{2a}{A}} B^j \tilde{\phi}_j(x) ,$$

where we define $z := y\sqrt{a + b + c}$, $\alpha := \frac{\sqrt{2c}}{\sqrt{a+b+c}}$ and use

$$(1 - \alpha^2)^{1/2} = \sqrt{1 - \frac{2c}{a+b+c}} = \sqrt{\frac{a + b - c}{a+b+c}} = \sqrt{\frac{(a + b)^2 - c^2}{(a+b+c)^2}} = \frac{b}{a+b+c} .$$

$\square$

Proof of Proposition 40

Since K is uniformly continuous, if d is the distance $E \times E$, there exists a δ_n such that

$$d((x_1, y_1), (x_2, y_2)) < \delta_n \Rightarrow |K(x_1, y_1) - K(x_2, y_2)| < n^{-1}$$

for $n = 1, 2, \ldots$ and arbitrary $x_1, x_2, y_1, y_2 \in E$. Since E is compact, we can cover it with a finite number of balls $\{E_{n,i}\}_{i=1}^m$ of diameter δ_n. If we arbitrarily choose $v_i \in E_{n,i}$ and define $K_n(x, y) := K(v_i, v_j)$ for $(x, y) \in E_{n,i} \times E_{n,j}$, from the uniform continuity of K, we obtain

$$\max_{(x,y)\in E\times E}|K(x,y)-K_n(x,y)|<\frac{1}{n}.$$

Let T_K, T_{K_n} be the integral operators of K, K_n. Then, we have

$$|\langle T_K f, f\rangle - \langle T_{K_n} f, f\rangle| \le n^{-1}\|f\|^2$$

and

$$\langle T_{K_n} f, f\rangle = \sum_{i=1}^{m}\sum_{j=1}^{m} K(v_i, v_j) \int_{E_{n,i}} f(x)d\mu(x) \int_{E_{n,j}} f(y)d\mu(y)$$

for an arbitrary n, and we have $\langle T_K f, f\rangle \ge 0$. Conversely, suppose that $\langle T_K f, f\rangle \ge 0$. If there exist $x_1, \ldots, x_m \in E$, $z_1, \ldots, z_m \in \mathbb{R}$ such that $\sum_{i=1}^{m}\sum_{j=1}^{m} z_i z_j k(x_i, x_j) < 0$, since K is uniformly continuous, there exist $E_1, \ldots, E_m \in \mathcal{F}$ such that

$$\max_{x_h, y_h \in E_h, h=1,\ldots,m} \sum_{i=1}^{m}\sum_{j=1}^{m} z_i z_j K(x_i.y_j) < 0$$

and $\mu(E_1), \ldots, \mu(E_m) > 0$. However, from the mean value theorem, we have

$$\langle T_K f, f\rangle := \sum_{i=1}^{m}\sum_{j=1}^{m} z_i z_j \{\mu(E_i)\mu(E_j)\}^{-1} \int_{E_i}\int_{E_j} k(x, y)d\mu(x)d\mu(y) < 0$$

for $f = \sum_{i=1}^{m} z_i \{\mu(E_i)\}^{-1} I_{E_i}$, which contradicts the fact that T_K is positive definite. $\qquad\square$

Proof of Lemma 3

We assume that $f_n(x)$ monotonically increases as n grows for each $x \in E$. Let $\epsilon > 0$ be arbitrary. For each $x \in E$, let $n(x)$ be the minimum n such that $|f_n(x) - f(x)| < \epsilon$. From continuity, for each $x \in E$, we set $U(x)$ so that

$$y \in U(x) \Longrightarrow |f(x) - f(y)| < \epsilon, \quad |f_{n(x)}(x) - f_{n(x)}(y)| < \epsilon.$$

Then, we have

$$f(y) - f_{n(x)}(y) \le f(x) + \epsilon - f_{n(x)}(y) \le f_{n(x)}(x) + 2\epsilon - f_{n(x)}(y) \le |f_{n(x)}(x) - f_{n(x)}(y)| + 2\epsilon < 3\epsilon.$$

Moreover, since E is compact, we may suppose that $E \subseteq \cup_{i=1}^{m} U(x_i)$. If N is the maximum value of $n(x_1), \ldots, n(x_m)$, for $n \ge N$, we have

$$f(y) - f_n(y) \le f(y) - f_{n(x_i)}(y) \le 3\epsilon$$

for each $y \in E$ and each i for which $y \in U(x_i)$. $\qquad\square$

Exercises 31~45

31. Proposition 34 can be derived according to the following steps. Which part of the proof in the appendix does each step correspond to?

    (a) Define the inner product $\langle \cdot, \cdot \rangle_{H_0}$ of $H_0 := \text{span}\{k(x, \cdot) : x \in E\}$.
    (b) For any Cauchy sequence $\{f_n\}$ in H_0 and each $x \in E$, the real sequence $\{f_n(x)\}$ is Cauchy, so it converges to $af(x) := \lim_{n \to \infty} f_n(x)$ (Proposition 6).

    Let H be such a set of fs.
    (c) Define the inner product $\langle \cdot, \cdot \rangle_H$ of the linear space H.
    (d) Show that H_0 is dense in H.
    (e) Show that any Cauchy sequence $\{f_n\}$ in H converges to some element of H as $n \to \infty$ (completeness of H).
    (f) Show that k is a reproducing kernel of H.
    (g) Show that such an H is unique.

32. In Examples 55 and 56, the inner product is $\langle f, g \rangle_H = \int_0^1 F(u)G(u)du$, and the RKHS is

$$ H = \{E \ni x \mapsto \int_E F(t)J(x, t)d\eta(t) \in \mathbb{R} | F \in L^2(E, \eta)\} \,. $$

What are the $J(x, t)$ in Examples 55 and 56? Also, how is the kernel $k(x, y)$ represented in general by using $J(x, t)$?

33. Proposition 38 can be derived according to the following steps. Which part of the proof in the appendix does each step correspond to?

    (a) Fix $f \in H$ arbitrarily define $N^{\perp} \ni (f_1, f_2) := v^{-1}(f), k(x, \cdot) := k_1(x, \cdot) + k_2(x, \cdot)$, and $(h_1(x, \cdot), h_2(x, \cdot)) := v^{-1}(k(x, \cdot))$, and show that

$$ \langle f_1, h_1(x, \cdot) \rangle_1 + \langle f_2, h_2(x, \cdot) \rangle_2 = \langle f_1, k_1(x, \cdot) \rangle_1 + \langle f_2, k_2(x, \cdot) \rangle_2. $$

    (b) Using (a), prove the reproducing property of k: $\langle f, k(x, \cdot) \rangle_H = f(x)$.
    (c) Show that the norm of H is (3.6)

34. Show that each $f \in W_q[0, 1]$ can be the Taylor series expanded by

$$ f(x) = \sum_{i=0}^{q-1} f^{(i)}(0)\phi_i(x) + \int_0^1 G_q(x, y)f^{(q)}(y)dy $$

using

$$ \phi_i(x) := \frac{x^i}{i!} , \quad i = 0, 1, \ldots $$

and

$$G_q(x, y) := \frac{(x - y)_+^{q-1}}{(q - 1)!} .$$

35. Show that $W_q[0, 1] = H_0 \oplus H_1$, where

$$H_0 = \{\sum_{i=0}^{q-1} \alpha_i \phi_i(x) | \alpha_0, \dots, \alpha_{q-1} \in \mathbb{R}\}$$

$$H_1 = \{\int_0^1 G_q(x, y)h(y)dy | h \in L^2[0, 1]\}$$

(You need to show the inclusion relation on both sides of the set). In addition, show that $H_0 \cap H_1 = \{0\}$.

36. We consider the integral operator T_k of $k(x, y) = \min\{x, y\}$, in $L^2[0, 1]$, where $x, y \in E = [0, 1]$. Substitute

$$\lambda_j = \frac{4}{\{(2j - 1)\pi\}^2}$$

$$e_j(x) = \sqrt{2} \sin\left(\frac{(2j - 1)\pi}{2}x\right)$$

into $T_k e_j = \lambda_j e_j$ to examine the equality.

37. Show that the eigenvalues in Example 59 form a geometric sequence with the initial values and ratio that are determined by $\beta := \hat{\sigma}^2/\sigma^2$.

38. In Example 59, the following program obtains eigenvalues and eigenfunctions under the assumption that $\sigma^2 = \hat{\sigma}^2 = 1$. We can change the program to set the values of $\sigma^2, \hat{\sigma}^2$ in ## and add $\sigma^2, \hat{\sigma}^2$ as an argument to the function phi in ### and run it to output a graph.

```
1  H=function(j,x) if(j==0) 1 else if(j==1) 2*x else if(j==2)-2+4*x^2
       else 4*x-8*x^3
2  cc=sqrt(5)/4; a=1/4                                           ##
3  phi=function(j,x) exp(-(cc-a)*x^2)*H(j,sqrt(2*cc)*x)    ###
4  curve(phi(0,x),-2,2, ylim=c(-2,8),col=1,ylab="phi")
5  for(i in 1:3)curve(phi(i,x),-2,2, ylim=c(-2,8), add=TRUE,
6      ann=FALSE, col=i+1)
7  legend("topright",legend=paste("j=",0:3),lwd=1, col=1:4)
8  title("Eigenfunction of Gaussian kernel")
```

39. Show the following:

(a) The function $f_n(x) = n^2(1 - x)x^{n+1}$ defined over $[0, 1]$ converges at each $x \in [0, 1]$, but its upper bound does not converge (it is not uniformly convergent).

(b) The function $f_n(x) = (1 - x)x^{n+1}$ defined over $[0, 1]$ converges uniformly (using Lemma 3).

(c) The series $\sum_{n=0}^{\infty} \frac{(-1)^n}{\sqrt{n+1}}$ converges absolutely.

40. In Example 58, suppose that the period of ϕ is 2π instead of 2. What are the eigenvalues and eigenfunctions of T_k? Additionally, derive the kernel k.

41. What eigenequations should be solved in Example 61 when $m = 3, d = 1$?

42. Define and execute the following part of the program in Example 62 as a function. The input for this includes data x, a kernel k, and the i of the i-th eigenvalue. The output is a function F.

```
1   K=matrix(0,m,m)
2   for(i in 1:m)for(j in 1:m)K[i,j]=k(x[i],x[j])
3     eig=eigen(K)
4     lam.m=eig$values
5     lam=lam.m/m
6     U=eig$vector
7     alpha=array(0,dim=c(m,m))
8     for(i in 1:m)alpha[,i]=U[,i]*sqrt(m)/lam.m[i]
9     F=function(y,i){
10    S=0; for(j in 1:m)S=S+alpha[j,i]*k(x[j],y)
11    return(S)
12  }
```

43. In Example 62, for the Gaussian kernel, random numbers are generated according to the normal distribution, and we obtain the corresponding eigenvalues and eigenfunctions. When the number of samples is large, theoretically, the eigenvalues are reduced exponentially (Example 59). What happens with the polynomial kernel $k(x, y) = (1 + xy)^2$ when $m = 2$ and $d = 1$? Output the eigenvalues and eigenfunctions as the Gaussian kernel.

44. If we construct (3.19) using the solution of $K_m U = U\Lambda$, show that the result is a solution of (3.18) and that it is orthogonal with a magnitude of 1.

45. In Proposition 42, β_j should originally satisfy $\sum_{j=1}^{\infty} \beta_j^2 < \infty$. However, this is not stated in the assertion of Proposition 42. Why is this the case?

Chapter 4
Kernel Computations

In Chap. 1, we learned that the kernel $k(x, y) \in \mathbb{R}$ represents the similarity between two elements x, y in a set E. Chapter 3 described the relationships between a kernel k, its feature map $E \ni x \mapsto k(x, \cdot) \in H$, and its reproducing kernel Hilbert space H. In this chapter, we consider $k(x, \cdot)$ to be a function of $E \to \mathbb{R}$ for each $x \in E$, and we perform data processing for N actual data pairs $(x_1, y_1), \ldots, (x_N, y_N)$ of covariates and responses. The x_i, $i = 1, \ldots, N$ (row vectors) are p-dimensional and given by the matrix $X \in \mathbb{R}^{N \times p}$. The responses y_i $(i = 1, \ldots, N)$ may be real or binary. This chapter discusses kernel ridge regression, principal component analysis, support vector machines (SVMs), and splines, and we find the $f \in H$ that minimizes the objective function under various constraints. It is known that we can write the optimal f in the form $\sum_{i=1}^{N} \alpha_i k(x_i, \cdot)$ (representation theorem), and the problem reduces to finding the optimal $\alpha_1, \ldots, \alpha_N$.

In the second half, we address the problem of computational complexity. The computation of a kernel takes more than $O(N^3)$, and real-time calculation is hard when N is greater than 1000. In particular, we consider how to reduce the rank of the Gram matrix K. Specifically, we learn actual procedures for random Fourier features, Nyström approximation, and incomplete Cholesky decomposition.

4.1 Kernel Ridge Regression

We say that finding the $\beta \in \mathbb{R}^p$ (column vector) that minimizes $\sum_{i=1}^{N}(y_i - x_i \beta)^2$ is the least squares problem. If we assume that we have executed the centralization process such that $y_i \leftarrow y_i - \bar{y}$ and $x_{i,j} \leftarrow x_{i,j} - \bar{x}_j$ for $\bar{y} = \dfrac{1}{N} \displaystyle\sum_{i=1}^{N} y_i$ and

© The Author(s), under exclusive license to Springer Nature Singapore Pte Ltd. 2022
J. Suzuki, *Kernel Methods for Machine Learning with Math and R*,
https://doi.org/10.1007/978-981-19-0398-4_4

$\bar{x}_j = \dfrac{1}{N} \displaystyle\sum_{i=1}^{N} x_{i,j}$ and that the matrix $X^\top X$ is nonsingular, we can obtain the solution

as $\hat{\beta} = (X^\top X)^{-1} X^\top y$ from $X = (x_{i,j})$ and $y = (y_i)$. In the following, we prepare a kernel $k : E \times E \to \mathbb{R}$ and consider the problem of finding the $f \in H$ that minimizes

$$L := \sum_{i=1}^{N} (y_i - f(x_i))^2$$

As we considered in Example 40, we express the RKHS H as the sum of

$$M := \mathrm{span}(\{k(x_i, \cdot)\}_{i=1}^{N})$$

and

$$M^\perp = \{f \in H \,|\, \langle f, k(x_i, \cdot) \rangle_H = 0, \ i = 1, \ldots, N\}.$$

If we set $f = f_1 + f_2$, $f_1 \in M$, $f_2 \in M^\perp$, then we have

$$\sum_{i=1}^{N} (y_i - f(x_i))^2 = \sum_{i=1}^{N} (y_i - f_1(x_i))^2 = \sum_{i=1}^{N} \left(y_i - \sum_{j=1}^{N} \alpha k(x_j, x_i)\right)^2 \tag{4.1}$$

and $E := \mathbb{R}^p$; we then obtain

$$f(x_i) = \langle f_1(\cdot) + f_2(\cdot), k(x_i, \cdot) \rangle_H = \langle f_1(\cdot), k(x_i, \cdot) \rangle_H = f_1(x_i)$$

for $i = 1, \ldots, N$. Thus, the minimization of L reduces to that of

$$L = \sum_{i=1}^{N} \{y_i - \sum_{j=1}^{N} \alpha_j k(x_j, x_i)\}^2 = \|y - K\alpha\|^2 \tag{4.2}$$

where $K = (k(x_i, x_j))_{i,j=1,\ldots,N}$ is a Gram matrix, and the norm $\|z\|$ of $z = [z_1, \ldots, z_N] \in \mathbb{R}$ denotes $\sqrt{\sum_{i=1}^{N} z_i^2}$. The above principle is the representation theorem.

If we differentiate L by α, we have $-K(y - K\alpha) = 0$. If K is positive definite rather than nonnegative definite, then the solution becomes $\hat{\alpha} = K^{-1} y$.

If we use the $\hat{f} \in H$ obtained as above that minimizes (4.2), then we can predict the value of y given a new $x \in \mathbb{R}^p$ via

$$\hat{f}(x) = \sum_{i=1}^{n} \hat{\alpha}_i k(x_i, x).$$

Fig. 4.1 We execute kernel
regression by using
polynomial and Gaussian
kernels

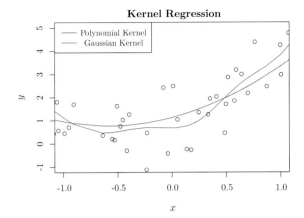

We can construct a procedure to compute α as follows.

```
1  alpha=function(k,x,y){
2    n=length(x); K=matrix(0,n,n)
3    for(i in 1:n)for(j in 1:n)K[i,j]=k(x[i],x[j])
4    return(solve(K+10^(-5)*diag(n))%*%y)    ## K might not be nonsingular
5  }
```

Example 63 Utilizing the function `alpha`, we execute kernel regression via polynomial and Gaussian kernels for $n = 50$ data ($\lambda = 0.1$). We present the output in
Fig. 4.1.

```
1   k.p=function(x,y) (sum(x*y)+1)^3        ## Kernek Definition
2   k.g=function(x,y) exp(-(x-y)^2/2)       ## Kernel Definition
3   lambda=0.1
4   n=50; x=rnorm(n); y=1+x+x^2+rnorm(n)         ## Data Generation
5   alpha.p=alpha(k.p,x,y); alpha.g=alpha(k.g,x,y)
6   z=sort(x); u=array(n); v=array(n)
7   for(j in 1:n){
8     S=0;for(i in 1:n)S=S+alpha.p[i]*k.p(x[i],z[j]); u[j]=S
9     S=0;for(i in 1:n)S=S+alpha.g[i]*k.g(x[i],z[j]); v[j]=S
10  }
11  plot(z,u,type="l",xlim=c(-1,1),xlab="x", ylab="y", ylim=c(-1,5),
12    col="red",main="Kernel Regression")
13  lines(z,v,col="blue"); points(x,y)
14  legend("topleft", legend = c("Polynomial kernel","Gaussian kernel"),
15    col = c("red","blue"), lty = 1)
```

We cannot obtain the solution of a linear regression problem when the rank of X
is smaller than p, i.e., $N < p$. Thus, we often minimize

$$\sum_{i=1}^{N}(y_i - x_i \beta)^2 + \lambda \|\beta\|_2^2$$

for cases in which $\lambda > 0$. We call such a modification of linear regression a ridge. The β to be minimized is given by $(X^\top X + \lambda I)^{-1} X^\top y$. In fact, we derive the formula by differentiating

$$\|y - X\beta\|^2 + \lambda \beta^\top \beta$$

by β and equating it to zero; we obtain

$$-X^\top(y - X\beta) + \lambda \beta = 0.$$

We consider extending ridge regression to the problem of finding the $f \in H$ that minimizes

$$L' := \sum_{i=1}^{N}(y_i - f(x_i))^2 + \lambda \|f\|_H^2. \tag{4.3}$$

Since f_1 and f_2 are orthogonal, we have

$$\|f\|_H^2 = \|f_1\|_H^2 + \|f_2\|_H^2 + 2\langle f_1, f_2 \rangle_H = \|f_1\|_H^2 + \|f_2\|_H^2 \geq \|f_1\|_H^2. \tag{4.4}$$

From (4.1), (4.3), and (4.4), we also have

$$L' \geq \sum_{i=1}^{N}(y_i - f_1(x_i))^2 + \lambda \|f_1\|_H^2.$$

If we note that the second term can be expressed by

$$\|f_1\|_H^2 = \langle \sum_{i=1}^{N} \alpha_i k(x_i, \cdot), \sum_{j=1}^{N} \alpha_j k(x_j, \cdot) \rangle_H = \sum_{i=1}^{N}\sum_{j=1}^{N} \alpha_i \alpha_j \langle k(x_i, \cdot), k(x_j, \cdot) \rangle_H = \alpha^\top K \alpha$$

for $\alpha = [\alpha_1, \ldots, \alpha_N]^\top$, then the minimization of L' reduces to that of

$$\|y - K\alpha\|^2 + \lambda \alpha^\top K \alpha. \tag{4.5}$$

If we differentiate the equation by α and set it equal to zero, we obtain

$$-K(y - K\alpha) + \lambda K\alpha = 0.$$

If K is nonsingular, we have

Fig. 4.2 We execute kernel ridge regression using polynomial and Gaussian kernels

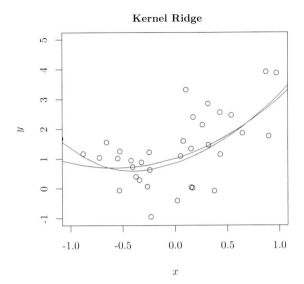

$$\hat{\alpha} = (K + \lambda I)^{-1} y. \tag{4.6}$$

Finally, if we use the $\hat{f} \in H$ that minimizes the (4.3) obtained thus far, we can predict the value of y given a new $x \in \mathbb{R}^p$ via

$$\hat{f}(x) = \sum_{i=1}^{n} \hat{\alpha}_i k(x_i, x).$$

For example, we can construct a procedure that finds α as follows.

```
1  alpha=function(k,x,y){
2      n=length(x); K=matrix(0,n,n)
3      for(i in 1:n)for(j in 1:n)K[i,j]=k(x[i],x[j])
4      return(solve(K+lambda*diag(n))%*%y)
5  }
```

Example 64 Using the function `alpha`, we execute kernel ridge regression for polynomial and Gaussian kernels and $n = 50$ data($\lambda = 0.1$). We show the outputs in Fig. 4.2.

```
1  k.p=function(x,y) (sum(x*y)+1)^3      ## Kernel Definition
2  k.g=function(x,y) exp(-(x-y)^2/2)     ## Kernel Definition
3  lambda=0.1
4  n=50; x=rnorm(n); y=1+x+x^2+rnorm(n)         ## Data Generation
5  alpha.p=alpha(k.p,x,y); alpha.g=alpha(k.g,x,y)
6  z=sort(x); u=array(n); v=array(n)
7  for(j in 1:n){
8      S=0;for(i in 1:n)S=S+alpha.p[i]*k.p(x[i],z[j]); u[j]=S
```

```
 9      S=0;for(i in 1:n)S=S+alpha.g[i]*k.g(x[i],z[j]); v[j]=S
10   }
11   plot(z,u,type="l",xlim=c(-1,1),xlab="x", ylab="y", ylim=c(-1,5),col
        ="red",main="Kernel Ridge")
12   lines(z,v,col="blue"); points(x,y)
```

4.2   Kernel Principle Component Analysis

We review the procedure of principal component analysis (PCA) when we do not use any kernel. We centralize each of the columns in the matrix X and vector y. We first compute the $v_1 := v \in \mathbb{R}^p$ that maximizes $v^\top X^\top X v$ under $v^\top v = 1$. Similarly, for $i = 2, \ldots, p$, we repeatedly compute v_i with the $v^\top v = 1$ that maximizes $v^\top X^\top X v$ and is orthogonal to $v_1, \cdots, v_{i-1} \in \mathbb{R}^p$. In the actual cases, we do not use all of the $v_1, \cdots, v_p$ but compress $\mathbb{R}^p$ to the $v_1, \cdots, v_m$ ($1 \leq m \leq p$) with the largest eigenvalues. We compute the $v \in \mathbb{R}^p$ that maximizes

$$v^\top X^\top X v - \mu(v^\top v - 1) \tag{4.7}$$

with a $\mu > 0$ Lagrange coefficient to find $v \in \mathbb{R}^p$ with the $v^\top v = 1$ that maximizes $v^\top X^\top X v$. In PCA, we often compute

$$\begin{bmatrix} x v_1 \\ \vdots \\ x v_m \end{bmatrix} \in \mathbb{R}^m$$

for each row vector $x \in \mathbb{R}^p$ using the obtained $v_1, \ldots, v_m \in \mathbb{R}^p$. We call such a value the score of x, which is the vector obtained by projecting x onto the m elements.

We may apply a problem that is similar to PCA for an RKHS H via the feature map $\Phi : E \ni x_i \mapsto k(x_i, \cdot) \in H$ rather than the PCA in $\mathbb{R}^p$. To this end, we consider the problem of finding the $f \in H$ that maximizes

$$\sum_{j=1}^N f(x_i)^2 - \mu(\|f\|_H^2 - 1), \tag{4.8}$$

with an $\mu > 0$ Lagrange coefficient.

If we use the linear kernel (the standard inner product), we can express $f \in H$ by $f(\cdot) = \langle w, \cdot \rangle_E$ with $w \in E$. Thus, (4.7) and (4.8) coincide. The centralization in the kernel PCA is for the Gram matrix K rather than the matrix X. For the other part, the extension follows in the same manner.

As discussed in the previous section, we apply the representation theorem. Thus, for $f_1 \in M := \mathrm{span}(\{k(x_i, \cdot)\}_{i=1}^N)$ and $f_2 \in M^\perp$, we have

$$\sum_{i=1}^{N} f(x_i)^2 = \sum_{i=1}^{N} \langle f_1(\cdot) + f_2(\cdot), k(x_i, \cdot) \rangle_H^2 = \sum_{i=1}^{N} \langle f_1(\cdot), k(x_i, \cdot) \rangle_H^2 = \sum_{i=1}^{N} f_1(x_i)^2$$

$$= \sum_{i=1}^{N} \{ \sum_{j=1}^{N} \alpha_j k(x_j, x_i) \}^2 = \sum_{i=1}^{N} \sum_{r=1}^{N} \sum_{s=1}^{N} \alpha_r \alpha_s k(x_r, x_i) k(x_s, x_i) = \alpha^\top K^2 \alpha$$

$$\| f_1 + f_2 \|_H^2 = \| f_1 \|_H^2 + \| f_2 \|_H^2 \geq \| f_1 \|_H^2$$

$$= \| \sum_{j=1}^{N} \alpha_j k(x_j, \cdot) \|_H^2 = \sum_{r=1}^{N} \sum_{s=1}^{N} \alpha_r \alpha_s k(x_r, x_s) = \alpha^\top K \alpha.$$

Hence, we can formulate (4.8) as the maximization of

$$\alpha^\top K^\top K \alpha - \mu(\alpha^\top K \alpha - 1).$$

If we substitute $\beta = K^{1/2}\alpha$, then since K is symmetric, we have

$$\beta^\top K \beta - \mu(\beta^\top \beta - 1).$$

Let $\lambda_1, \ldots, \lambda_N$ and $u_1, \ldots, u_N$ be the eigenvalues and eigenvectors of the eigenequation $K\beta = \lambda\beta$, respectively. Then, we have [26]

$$\alpha = K^{-1/2}\beta = \frac{1}{\sqrt{\lambda}}\beta = \frac{u_1}{\sqrt{\lambda_1}}, \ldots, \frac{u_N}{\sqrt{\lambda_N}}.$$

If we centralize the Gram matrix $K = (k(x_i, x_j))$, then the (i, j)−th element of the modified Gram matrix is

$$\langle k(x_i, \cdot) - \frac{1}{N} \sum_{h=1}^{N} k(x_h, \cdot), k(x_j, \cdot) - \frac{1}{N} \sum_{h=1}^{N} k(x_h, \cdot) \rangle_H$$

$$= k(x_i, x_j) - \frac{1}{N} \sum_{h=1}^{N} k(x_i, x_h) - \frac{1}{N} \sum_{l=1}^{N} k(x_j, x_l)$$

$$+ \frac{1}{N^2} \sum_{h=1}^{N} \sum_{l=1}^{N} k(x_h, x_l). \tag{4.9}$$

To obtain the score (size $1 \leq m \leq p$) of $x \in \mathbb{R}^p$ (row vector), we use the first m columns of $A = [\alpha_1, \ldots, \alpha_N]^\top \in \mathbb{R}^{N \times p}$. Let $x_i \in \mathbb{R}^p$ and $\alpha_i \in \mathbb{R}^m$ be a row vector of X and the i−th column of $A \in \mathbb{R}^{N \times m}$, respectively. Then,

$$\sum_{i=1}^{N} \alpha_i k(x_i, x) \in \mathbb{R}^m$$

is the score of $x \in \mathbb{R}^p$.

Compared to ordinary PCA, kernel PCA requires a computational time of $O(N^3)$. Therefore, when N is large compared to p, the computational complexity may be enormous. In the R language, we can write the procedure as follows.

```
kernel.pca.train=function(x,k){
  n=nrow(x); K=matrix(0,n,n); S=rep(0,n); T=rep(0,n)
  for(i in 1:n)for(j in 1:n)K[i,j]=k(x[i,],x[j,])
  for(i in 1:n)S[i]=sum(K[i,])
  for(j in 1:n)T[j]=sum(K[,j])
  U=sum(K)
  for(i in 1:n)for(j in 1:n)K[i,j]=K[i,j]-S[i]/n-T[j]/n+U/n^2
  res=eigen(K)
  alpha=matrix(0,n,n)
  for (i in 1:n)alpha[,i]=res$vector[,i]/res$value[i]^0.5
  return(alpha)
}

kernel.pca.test=function(x,k,alpha,m,z){
  n=nrow(x)
  pca=array(0,dim=m)
  for(i in 1:n)pca=pca+alpha[i,1:m]*k(x[i,],z)
  return(pca)
}
```

In kernel PCA, when we use the linear kernel, the scores are consistent with those of PCA without any kernel. For simplicity, we assume that X is normalized. If we do not use the kernel, then by the singular value decomposition of $X = U \Sigma V^\top$ ($U \in \mathbb{R}^{N \times p}$, $\Sigma \in \mathbb{R}^{p \times p}$, $V \in \mathbb{R}^{p \times p}$), the multiplication of $\frac{1}{N-1} X^\top X = \frac{1}{N-1} V \Sigma^2 V^\top$ and $V^\top$ is $\frac{1}{N-1} X^\top X V = \frac{1}{N-1} V \Sigma^2$. Thus, each column of V is a principal component vector, and the scores of $x_1, \ldots, x_N \in \mathbb{R}^p$ (row vector) are the first m columns of

$$XV = U \Sigma V^\top \cdot V = U \Sigma.$$

On the other hand, for the linear kernel, we may write the Gram matrix as $K = XX^\top = U \Sigma^2 U^\top$ and have $KU = XX^\top U = U \Sigma^2$. That is, each column of U is $\beta_1, \ldots, \beta_N$, and the columns $\alpha_1, \ldots, \alpha_N$ of $K^{-1/2} U$ are the principal component vectors. Therefore, the scores of $x_1, \ldots, x_N \in \mathbb{R}^p$ (row vectors) are the first m columns of

$$K \cdot K^{-1/2} U = U \Sigma^2 U^\top \cdot (U \Sigma^2 U^\top)^{-1/2} \cdot U = U \Sigma.$$

Furthermore, we compare the results in terms of centralization. Equation (4.9) is

$$x_i x_j - \frac{1}{N} \sum_{h=1}^{N} x_i x_h - \frac{1}{N} \sum_{l=1}^{N} x_j x_l + \frac{1}{N} \sum_{h=1}^{N} \sum_{l=1}^{N} x_l x_h = (x_i - \bar{x})(x_j - \bar{x})$$

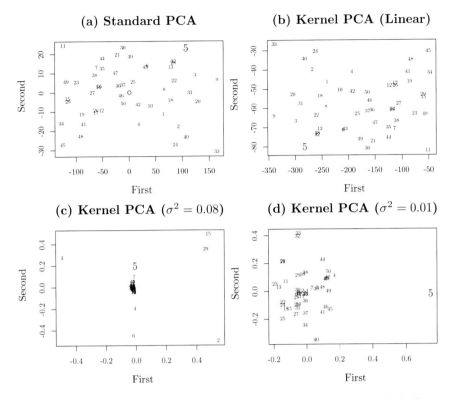

Fig. 4.3 For the US Arrests data, we ran the ordinary PCA and kernel PCA methods (linear; Gaussian with $\sigma^2 = 0.08, 0.01$), and we display the scores here. In the figure, 1–50 are the IDs given to the states, and California's ID is 5 (written in red). The results of the kernel PCA approach differ greatly depending on what kernel we choose. Additionally, since kernel PCA is unsupervised, it is not possible to use CV to select the optimal parameters. The scores of ordinary PCA and PCA with the linear kernel should be identical. Although the directions of both axes are opposite, which is common in PCA, we can conclude that they match

for the linear kernel, which is consistent with that of the ordinary PCA approach. Therefore, the obtained score is the same.

Example 65 We performed kernel PCA on a data set called US Arrests in R. We wished to project the ratio of the resident population living in urban areas and the incidence rates of homicide, violent crime, and assaults on women (number of arrests per 100,000 people) for all 50 states of the U.S. onto the axes of two variables using PCA. We performed kernel PCA with a Gaussian kernel ($\sigma^2 = 0.01, 0.08$), kernel PCA with a linear kernel, and ordinary PCA. We observed that the differences in the features of the 50 states were not evident in the results of ordinary PCA and kernel PCA with the linear kernel (Fig. 4.3a,b). With the Gaussian kernel ($\sigma^2 = 0.08$), the 50 states were divided into four categories (Fig. 4.3c). As far as the data were concerned, California's figures (fewer homicides for a higher urban population) differed from those of the other states. Nevertheless, when we set $\sigma = 0.01$, the differences between California and the other 49 states became clear (Fig. 4.3d). We used the following code for the execution of the compared approaches.

```
1  #k=function(x,y) sum(x*y)
2  sigma.2=0.01; k=function(x,y) exp(-norm(x-y,"2")^2/2/sigma.2)
3  x=as.matrix(USArrests); n=nrow(x); p=ncol(x)
4  alpha=kernel.pca.train(x,k)
5  z=array(dim=c(n,2)); for(i in 1:n)z[i,]=kernel.pca.test(x,k,alpha,2,
      x[i,])
6  min.1=min(z[,1]); min.2=min(z[,2]); max.1=max(z[,1]); max.2=max(z
      [,2])
7  plot(0, xlim=c(min.1,max.1),ylim=c(min.2,max.2),xlab="First",ylab="
      Second",cex.lab=0.75,cex.axis = 0.75, main="Kernel PCA (Gauss
      0.01)")
8  for(i in 1:n)if(i!=5)text(z[i,1],z[i,2],labels=i,cex = 0.5)
9  text(z[5,1],z[5,2],5,col="red")
10 ## For Usual PCA, we can compute the score in one line.
11 z=prcomp(x)$x[,1:2]
```

4.3   Kernel SVM

Consider binary discrimination using support vector machines (SVMs). Given $X \in \mathbb{R}^{N \times p}$ and $y \in \{1, -1\}^N$, we find the boundary $Y = X\beta + \beta_0$ with the $\beta \in \mathbb{R}^p$ and $\beta_0 \in \mathbb{R}$ that maximize the margin. Let $\gamma \geq 0$. We wish to maximize the margin M by ranging $(\beta_0, \beta) \in \mathbb{R} \times \mathbb{R}^p$ and $\epsilon_i \geq 0, i = 1, \ldots, N$ to satisfy

$$\sum_{i=1}^{N} \epsilon_i \leq \gamma$$

and

$$y_i (\beta_0 + x_i \beta) \geq M (1 - \epsilon_i) , \quad i = 1, \ldots, N.$$

We often formulate this as the problem of minimizing

$$\frac{1}{2} \|\beta\|^2 + C \sum_{i=1}^{N} \epsilon_i \tag{4.10}$$

under $y_i (x_i \beta + \beta_0) \geq 1 - \epsilon_i$, $\epsilon_i \geq 0$ for $i = 1, \ldots, N$ by using a constant $C > 0$ (the prime problem). We further transform it into the problem of finding $0 \leq \alpha_i \leq C$, $i = 1, 2, \ldots, N$ that maximizes

$$\sum_{i=1}^{N} \alpha_i - \frac{1}{2} \sum_{i=1}^{N} \sum_{j=1}^{N} \alpha_i \alpha_j y_i y_j x_i x_j^\top \qquad (4.11)$$

under $\sum_{i=1}^{N} \alpha_i y_i = 0$, where x_i is the i-th row vector of X (the dual problem).[1] The constant $C > 0$ is a parameter that represents the flexibility of the boundary surface. The higher the value is, the more samples are used to determine the boundary (samples with $\alpha_i \neq 0$, i.e., support vectors). Although we sacrifice the fit of the data, we reduce the boundary variation caused by sample data to prevent overtraining. Then, from the support vectors, we can calculate the slope of the boundary with the following formula:

$$\beta = \sum_{i=1}^{N} \alpha_i y_i x_i^\top \in \mathbb{R}^p.$$

Then, suppose that we replace the boundary surface with a curved surface by replacing the inner product $x_i x_j^\top$ with a general nonlinear kernel $k(x_i, x_j)$. Then, we can obtain complicated boundary surfaces rather than planes. However, the theoretical basis for replacing the product with a kernel is not clear.

Therefore, in the following, we derive the same results by formulating the optimization using $k : E \times E \to \mathbb{R}$. As in to the previous application of the representation theorem, we find the $f \in H$ that minimizes

$$\frac{1}{2} \|f\|_H^2 + C \sum_{i=1}^{N} \epsilon_i - \sum_{i=1}^{N} \alpha_i [y_i \{f(x_i) + \beta_0\} - (1 - \epsilon_i)] - \sum_{i=1}^{N} \mu_i \epsilon_i. \qquad (4.12)$$

Noting that $f(x_i) = f_1(x_i)$, $i = 1, \ldots, N$ and $\|f\|_H \geq \|f_1\|_H$, we find $\gamma_1, \ldots, \gamma_N$ such that $f(\cdot) = \sum_{i=1}^{N} \gamma_i k(x_i, \cdot)$.

The Karush-Kuhn-Tucker (KKT) condition results in the following nine equations:

$$y_i \{f(x_i) + \beta_0\} - (1 - \epsilon_i) \geq 0$$

$$\epsilon_i \geq 0$$

$$\alpha_i [y_i \{f(x_i) + \beta_0\} - (1 - \epsilon_i)] = 0$$

$$\mu_i \epsilon_i = 0$$

$$\sum_j \gamma_j k(x_i, x_j) - \sum_j \alpha_j y_j k(x_i, x_j) = 0 \qquad (4.13)$$

---

[1] We see this derivation in several references, such as Joe Suzuki, "Statistical Learning with Math and R" (Springer); C. M. Bishop, "Pattern Recognition and Machine Learning," (Springer); Hastie, Tibshirani, and Fridman, "Elements of Statistical Learning" (Springer); and other primary machine learning books.

Fig. 4.4 After generating samples, we draw linear (planar) and nonlinear (curved) boundaries with support vector machines

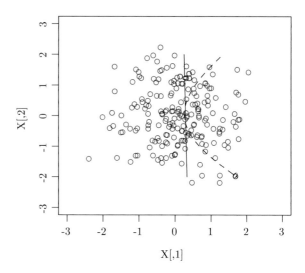

$$\sum_i \alpha_i y_i = 0$$

$$C - \alpha_i - \mu_i = 0 \qquad (4.14)$$

$$\mu_i \geq 0 , \ 0 \leq \alpha_i \leq C.$$

Next, suppose that $f_0, f_1, \ldots, f_m : \mathbb{R}^p \to \mathbb{R}$ are convex and differentiable at $\beta = \beta^*$. In general, the following Eqs. (4.15), (4.16) and (4.17) are called the KKT condition.[2]

Proposition 43 (KKT Condition) *Suppose that* $f_1(\beta) \leq 0, \ldots, f_m(\beta) \leq 0$. *Then,* $\beta = \beta^* \in \mathbb{R}^p$ *minimizes* $f_0(\beta)$ *if and only if*

$$f_1(\beta^*), \ldots, f_m(\beta^*) \leq 0 \qquad (4.15)$$

and $\alpha_1, \ldots, \alpha_m \geq 0$ *exist such that*

$$\alpha_1 f_1(\beta^*) = \cdots = \alpha_m f_m(\beta^*) = 0 \qquad (4.16)$$

$$\nabla f_0(\beta^*) + \sum_{i=1}^{m} \alpha_i \nabla f_i(\beta^*) = 0. \qquad (4.17)$$

Utilizing these nine equations, from (4.13) and (4.14), we can express (4.12) as

---

[2] For the proof, see Chap. 9 of Joe Suzuki "Statistical Learning with R/Python" (Springer).

$$\sum_{i=1}^{N} \alpha_i - \frac{1}{2} \sum_{i=1}^{N} \sum_{j=1}^{N} \alpha_i \alpha_j y_i y_j k(x_i . x_j). \qquad (4.18)$$

Comparing (4.11) and (4.18), we observe that the dual problem replaces $x_i^\top x_j$ with $k(x_i, x_j)$ for the formulation without any kernel.

In fact, if we set $f(\cdot) = \langle \beta, \cdot \rangle_H$, $\beta \in \mathbb{R}^p$, $k(x, y) = x^\top y$ $(x, y \in \mathbb{R}^p)$, then we obtain the dual problem for a linear kernel (4.11).

Example 66 By using the following function svm.2, we can compare how the bounds differ between a linear kernel (the standard inner product) and a nonlinear kernel (a polynomial kernel), as in Fig. 4.4. quadprog is an R package for solving quadratic programming problems. The function solve. QP calculates α.

```
library(quadprog)
K.linear <-function(x,y) {return(t(x)%*%y)}
K.poly <-function(x,y) {return((1+t(x)%*%y)^2)}
svm.2=function(X,y,C, K){      ## Function Name: svm.2
  eps=0.0001; n=nrow(X); Dmat=matrix(nrow=n,ncol=n);Kmat = matrix(
      nrow=n,ncol = n)
  for(i in 1:n)for(j in 1:n) {Dmat[i,j]=K(X[i,],X[j,])*y[i]*y[j];Kmat
      = K(X[i,],X[j,])}
  Dmat=Dmat+eps*diag(n); dvec=rep(1,n)
  Amat=matrix(nrow=(2*n+1),ncol=n); Amat[1,]=y; Amat[2:(n+1),1:n]=-
      diag(n);
  Amat[(n+2):(2*n+1),1:n]=diag(n) ; Amat=t(Amat)
  bvec=c(0,-C*rep(1,n),rep(0,n)); meq=1
  alpha=solve.QP(Dmat,dvec,Amat,bvec=bvec,meq=1)$solution
  index=(1:n)[0<alpha&alpha<C]
  beta=drop(Kmat%*%(alpha*y)); beta.0=mean(y[index]-beta[index])
  return(list(alpha=alpha,beta.0=beta.0))
}
# Function Definition
plot.kernel=function(K, lty){ ## The argument "lty" specifies the line type
  qq=svm.2(X,y,0.1,K); alpha=qq$alpha; beta.0=qq$beta.0
  f=function(u,v){x=c(u,v); S=beta.0; for(i in 1:n)S=S+alpha[i]*y[i]*
      K(X[i,], x); return(S)}
  ## f(x,y) specifies the height of z. We compute the contour from this
  u=seq(-2,2,.1);v=seq(-2,2,.1);w=array(dim=c(41,41))
  for(i in 1:41)for(j in 1:41)w[i,j]=f(u[i],v[j])
  contour(u,v,w,level=0,add=TRUE,lty=lty)
}
  # Execution
a=rnorm(1); b=rnorm(1)
n=100; X=matrix(rnorm(n*2),ncol=2,nrow=n); y=sign(a*X[,1]+b*X
    [,2]+0.3*rnorm(n))
plot(-3:3,-3:3,xlab="X[,1]",ylab="X[,2]", type="n")
for(i in 1:n){
  if(y[i]==1)points(X[i,1],X[i,2],col="red") else points(X[i,1],X[i
      ,2],col= "blue")
}
plot.kernel(K.linear,1); plot.kernel(K.poly,2)
```

4.4   Spline Curves

Let $J \geq 1$. We say that the function

$$g(x) = \beta_1 + \beta_2 x + \beta_3 x^2 + \beta_4 x^3 + \sum_{j=1}^{J} \beta_{j+4}(x - \xi_j)_+^3 \qquad (4.19)$$

$$= \begin{cases} g_0(x) = \beta_1 + \beta_2 x + \beta_3 x^2 + \beta_4 x^3, & x < \xi_1 \\ g_j(x) = g_{j-1}(x) + \beta_{j+4}(x - \xi_j)^3, & \xi_j \leq x < \xi_{j+1} \\ g_J(x) = \beta_1 + \beta_2 x + \beta_3 x^2 + \beta_4 x^3 + \sum_{j=1}^{J} \beta_{j+4}(x - \xi_j)^3, & x \geq \xi_J \end{cases}$$

with the constants $\beta_1, \ldots, \beta_{J+4} \in \mathbb{R}$ is a spline function of order three with knots $0 < \xi_1 < \cdots < \xi_J < 1$. We may define the spline function of order three by the function g, which is a piecewise polynomial for each of the $J + 1$ intervals whose g, g', g'' are continuous at the J knots. The spline expressed by (4.19) consists of a linear space, and

$$1, x, x^2, x^3, (x - \xi_1)_+^3, \ldots, (x - \xi_J)_+^3 \qquad (4.20)$$

can be its basis.

In particular, we consider the natural spline of order three in which we pose more conditions such as

$$g''(\xi_1) = g'''(\xi_1) = 0 \qquad (4.21)$$

and

$$g''(\xi_J) = g'''(\xi_J) = 0. \qquad (4.22)$$

The resulting curve is not of order three in $x \leq \xi_1$, $\xi_J \leq x$, and we approximate it by a line. The linear space of natural splines possesses J dimensions. In fact, from (4.21), we have

$$g'''(\xi_1) = 6\beta_4 = 0$$

$$g''(\xi_1) = 2\beta_3 + 6\beta_4 \xi_1 = 0 \iff \beta_3 = \beta_4 = 0.$$

Additionally, from (4.22), we have

$$g'''(\xi_J) = 6\beta_4 + 6\sum_{j=1}^{J} \beta_{j+4} = 0$$

$$g''(\xi_J) = 2\beta_3 + 6\beta_4\xi_J + 6\sum_{j=1}^{J}\beta_{j+4}(\xi_J - \xi_j) = 0$$

$$\Longleftrightarrow \sum_{j=1}^{J}\beta_{j+4} = \sum_{j=1}^{J}\beta_{j+4}\xi_j = 0.$$

Thus, the β_{J+3}, β_{J+4} values are determined by the other β_j; $j = 1, 2, 5, \ldots, J + 2$.

In the following, we consider the problem of finding the $f : [0, 1] \rightarrow \mathbb{R}$ that minimizes

$$\sum_{i=1}^{N}\{y_i - f(x_i)\}^2 + \lambda \int_0^1 \{f''(x)\}^2 dx \qquad (4.23)$$

given samples $(x_1, y_1), \ldots, (x_N, y_N) \in \mathbb{R} \times \mathbb{R}$. The second term is zero if the function is a straight line, but it becomes a significant value if the function deviates from a straight line. In other words, this term represents the complexity of the function f. The constant $\lambda \geq 0$ balances the two terms, and if it is large, the curve is smooth; if the constant is small, the curve follows the sample closely. Note that in general, the bounds $\xi_1, \ldots, \xi_J$ and $x_1, \ldots, x_N$ are defined separately.

In this case, it is known that the f that minimizes (4.23) is a natural spline of order three such that $f(x_i) = y_i$, $i = 1, \ldots, N$ at the N boundaries $\xi_1 = x_1, \ldots, \xi_N = x_N$.[3] However, f is once differentiable everywhere and twice differentiable almost everywhere with $\int_0^1 \{f''(x)\}^2 dx < \infty$, which implies that f is an element of $W_2[0, 1]$. A similar proposition holds for the general $W_q[0, 1]$.

Example 67 In the case of a natural spline with $q = 2$, if we choose the basis $g_1, \ldots, g_N$ appropriately, such as $g(\cdot) = \sum_{j=1}^{N}\beta_j g_j(\cdot)$, and $G = (\int_0^1 g_i^{(q)}(x)g_j^{(q)}(x)dx) \in \mathbb{R}^{N \times N}$, $y = [y_1, \ldots, y_N]$, then we obtain the optimal

$$[\beta_1, \ldots, \beta_N]^\top = (X^\top X + \lambda G)^{-1}X^\top y.$$

Figure 4.5 shows the graphs obtained for $\lambda = 1, 30, 80$.

```
1   ## d,h contruct the basis functions
2   d=function(j,x,knots){
3       K=length(knots);
4       (max((x-knots[j])^3,0)-max((x-knots[K])^3,0))/(knots[K]-knots[
        j])
5   }
6   h=function(j,x,knots){
7       K=length(knots);
8       if(j==1) return(1) else if(j==2)return(x) else return(d(j-2,x,
        knots)-d(K-1,x,knots))
9   }
10  ## G is the value obtained by integrating the twice differentiated function
```

---

[3] See Chap. 7 of this series ("Statistical Learning with R/Python" (Springer)) for the proof.

Fig. 4.5 Instead of giving
knots or the number of knots
in the smoothing spline, we
specify the λ value, which
indicates smoothness.
Comparing λ = 1, 30, 80, as
we increase the value of λ,
the spline does not follow the
observed data, but it
becomes smoother

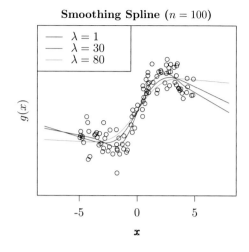

```
11   G=function(x){              ## Assume that each value of x is arranged ascending
12       n=length(x); g=matrix(0, nrow=n,ncol=n)
13       for(i in 3:(n-1))for(j in i:n){
14           g[i,j]=12*(x[n]-x[n-1])*(x[n-1]-x[j-2])*(x[n-1]-x[i-2])/(x[
             n]-x[i-2])/(x[n]-x[j-2])+
15           (12*x[n-1]+6*x[j-2]-18*x[i-2])*(x[n-1]-x[j-2])^2/(x[n]-x[i
             -2])/(x[n]-x[j-2])
16           g[j,i]=g[i,j]
17       }
18   return(g)
19   }
20   ## Main
21   n=100; x=runif(n,-5,5); y=x+sin(x)*2+rnorm(n)    ## Data Generation
22   index=order(x); x=x[index];y=y[index]
23   X=matrix(nrow=n,ncol=n); X[,1]=1
24   for(j in 2:n)for(i in 1:n)X[i,j]=h(j,x[i],x)   ## Generation of Matrix X
25   GG=G(x);                                       ## Generation of Matrix G
26   lambda.set=c(1,30,80); col.set=c("red","blue","green")
27   for(i in 1:3){
28       lambda=lambda.set[i]
29       gamma=solve(t(X)%*%X+lambda*GG)%*%t(X)%*%y
30       g=function(u){S=gamma[1]; for(j in 2:n)S=S+gamma[j]*h(j,u,x);
         return(S)}
31       u.seq=seq(-8,8,0.02); v.seq=NULL; for(u in u.seq)v.seq=c(v.seq,
         g(u))
32       plot(u.seq,v.seq,type="l",yaxt="n", xlab="x",ylab="g(x)",ylim=c
         (-8,8), col=col.set[i])
33       par(new=TRUE)
34   }
35   points(x,y); legend("topleft", paste0("lambda=",lambda.set), col=col
         .set, lty=1)
36   title("Smoothing Spline (n=100)")
```

Generalizing (4.23), we consider minimizing

$$\sum_{i=1}^{N} \{y_i - f(x_i)\}^2 + \lambda \int_0^1 \{f^{(q)}(x)\}^2 dx. \tag{4.24}$$

First, each $f = f_0 + f_1$ ($f_0 \in H_0$ and $f_1 \in H_1$ in $W_q[0, 1]$) can be written with the appropriate linear operators $P_0 \in B(H, H_0)$ and $P_1 \in B(H, H_1)$ as $f_0 = P_0 f \in H_0$, $f_1 = P_1 f \in H_1$. Since $\langle f_0, f_1 \rangle_H = 0$, f_0, f_1 minimize $\|f - f_0\|_H$ and $\|f - f_1\|_H$, respectively. Furthermore, P_0, P_1 are self-adjoint. In fact, from Proposition 19, for each $i = 0, 1$, we have

$$\langle P_i f, g \rangle_H = \langle f_i, g_0 + g_1 \rangle_H = \langle f_i, g_i \rangle_H = \langle f_0 + f_1, g_i \rangle_H = \langle f, P_i g \rangle_H$$

for $f_0, g_0 \in H_0$, $f_1, g_1 \in H_1$, $f = f_0 + f_1$, $g = g_0 + g_1$. Moreover, we have $P_i f \in H_i$ and $P_i^2 f = P_i f$. Thus, we can write the norm of the second term in (4.24) as

$$\int_0^1 |f^{(q)}(x)|^2 dx = \|P_1 f\|_{H_1}^2 = \langle P_1 f, P_1 f \rangle_{H_1} = \langle f, P_1^2 f \rangle_H = \langle f, P_1 f \rangle_H$$

and can express (4.24) as

$$\sum_{i=1}^{N} \{y_i - f(x_i)\}^2 + \lambda \langle f, P_1 f \rangle_H \tag{4.25}$$

for $f \in W_q[0, 1]$. Let $f = g + h \in H$, $g \in M := \text{span}\{\phi_0(\cdot), \ldots, \phi_{q-1}(\cdot), k(x_1, \cdot), \ldots k(x_N, \cdot)\}$, and $h \in M^\perp$. Then, for $i = 1, \ldots, N$, we have

$$f(x_i) = \langle g + h, k(x_i, \cdot) \rangle_H = g(x_i)$$

$$\|P_1 f\|_{H_1} \geq \|P_1 g\|_{H_1}$$

(the representation theorem). Thus, we may restrict the range of f to M for searching the optimum to find $\alpha_1, \ldots, \alpha_N, \beta_1, \ldots, \beta_q$ in

$$g(\cdot) = \sum_{i=0}^{q-1} \beta_i \phi_i(\cdot) + \sum_{i=1}^{N} \alpha_i k(x_i, \cdot). \tag{4.26}$$

In natural spline functions, we regard the differential of order q at $x = x_N$ as zero, which means that

$$g^{(q)}(x_N) = \ldots = g^{(2q-1)}(x_N) = 0, \tag{4.27}$$

and the dimensionality of $\text{span}\{k_1(x_i, \cdot) | i = 1, \ldots, N\}$ is $N - q$. For spline functions of order three ($q = 2$), (4.27) corresponds to (4.22). The basis $\{1, x\}$ w.r.t. the lines in $x \leq x_1$ corresponds to $\{\phi_0(x), \dots, \phi_{q-1}\}$. Thus, we find the optimal solution in the subspace of $W_q[0, 1]$ for N.

Proposition 44 *Let $r \in W_q[0, 1]$ be a natural spline with knots $x_1, \dots, x_N$ and a maximum order of $2q - 1$, and suppose that $g \in W_q[0, 1]$ satisfies $g(x_i) = r(x_i)$ for $i = 1, 2, \dots, N$. Then, we have*

$$\int_0^1 \{r^{(q)}(x)\}^2 dx \leq \int_0^1 \{g^{(q)}(x)\}^2 dx$$

Moreover, the maximum order of s is $q - 1$ such that $s(x_i) = 0$ for $s := g - r$ and $i = 1, 2, \dots, N$, and if $N \geq q$, then the function s is zero.

Proof: See the Appendix at the end of this chapter.

Since the natural splines of the highest order $2q - 1$ possess N dimensions, an $r \in W_q[0, 1]$ exists that shares the values $r(x_1) = g(x_1), \dots, r(x_N) = g(x_N)$ at the N boundaries $x_1, \dots, x_N$. Among them, since the second term in (4.25) is the optimum, the natural spline of the highest order $2q - 1$ is optimal.

To summarize the above, the problem of finding the f that minimizes (4.25) in $W_q[0, 1]$ reduces to finding the solution over the range of (4.26), (4.27). In other words, we can think of the problem in a subspace with N dimensions.

Moreover, the basis consists of N elements regardless of whether $q \geq 1$, and if we set $g(\cdot) = \sum_{j=1}^N \beta_j g_j(\cdot)$, the problem is to find the $\beta_1, \dots, \beta_N$ that minimize

$$\sum_{i=1}^N \{y_i - \sum_{i=1}^n \sum_{j=1}^N \beta_j g_j(x_i)\}^2 + \lambda \sum_{i=1}^N \sum_{j=1}^N \beta_i \beta_j \int_0^1 g_i^{(q)}(x) g_j^{(q)}(x) dx.$$

Let $X = (g_j(x_i)) \in \mathbb{R}^{N \times N}$, $G = (\int_0^1 g_i^{(q)}(x) g_j^{(q)}(x) dx) \in \mathbb{R}^{N \times N}$, and $y = [y_1, \dots, y_N]^\top$. The optimal solution $\beta = [\beta_1, \dots, \beta_N]^\top$ is given by

$$\beta = (X^\top X + \lambda G)^{-1} X^\top y.$$

4.5   Random Fourier Features

In the following, we examine computational cost reduction methods.

In particular, in this section, we learn about random Fourier features, which we can apply to the case where the kernel $k(x, y)$ $(x, y \in E)$ is a function of $x - y$.

Proposition 45 (Rahimi and Recht [23]) *Suppose that $k : E \times E \ni (x, y) \mapsto k(x, y) \in \mathbb{R}$ is a function of $x - y$. Then, we have*

$$k(x, y) = 2\mathbb{E}_{\omega, b} \cos(w^\top x + b) \cos(w^\top y + b) \tag{4.28}$$

where the expectation $\mathbb{E}_{\omega,b}$ is calculated over $\omega \sim \mu$ (the probability of k in Proposition 5) and $b \in [0, 2\pi)$ (the uniform distribution).

Proof: The claim is due to Bochner's theorem (Proposition 5). See the Appendix at the end of this chapter for details.

Based on Proposition 45, we generate $\sqrt{2}\cos(\omega^\top x + b)$ $m \geq 1$ times, i.e., (w_i, b_i), $i = 1, \ldots, m$, and construct the function

$$z_i(x) = \sqrt{2}\cos(\omega_i^\top x + b_i) \ i = 1, \ldots, m.$$

From the law of large numbers, the constructed

$$\hat{k}(x, y) := \frac{1}{m} \sum_{i=1}^{m} z_i(x) z_i(y)$$

approaches $k(x, y)$. Utilizing this fact, when m is small compared to N, the method to reduce the complexity of kernel computation is called random Fourier features (RFF).

We claim that the RFF possesses the following property.

$$P(|k(x, y) - \hat{k}(x, y)| \geq \epsilon) \leq 2\exp(-m\epsilon^2/8). \tag{4.29}$$

Proposition 46 (Hoeffding's Inequality) *For independent random variables X_i, $i = 1, \ldots, n$, each of which takes values in $[a_i, b_i]$, and an arbitrary $\epsilon > 0$, we have*

$$P(|\overline{X} - \mathbb{E}[\overline{X}]| \geq \epsilon) \leq 2\exp(-\frac{2n^2\epsilon^2}{\sum_{i=1}^{n}(b_i - a_i)^2}) \tag{4.30}$$

where $\overline{X}$ denotes the sample mean $(X_1 + \ldots + X_n)/n$.

Proof: We use the Chernoff bound and Hoeffding's lemma, which are shown below.

Lemma 5 (Chernoff Bound) *For a random variable X and an arbitrary $\epsilon > 0$, we have*

$$P(X \geq \epsilon) \leq \inf_{s>0} e^{-s\epsilon} \mathbb{E}[e^{sX}]. \tag{4.31}$$

To prove this lemma, we use the following lemma.

Lemma 6 (Markov's Inequality) *For a random variable X that takes nonnegative values, we have*

$$P(X \geq \epsilon) \leq \frac{\mathbb{E}[X]}{\epsilon}.$$

Lemma 6 is due to

$$\mathbb{E}[X] = \mathbb{E}[X \cdot I(X \geq \epsilon)] + \mathbb{E}[X \cdot I(X < \epsilon)] \geq \mathbb{E}[X \cdot I(X \geq \epsilon)] \geq \epsilon P(X \geq \epsilon).$$

Lemma 5 follows from lemma 6 and the fact that

$$P(X \geq \epsilon) = P(sX \geq s\epsilon) = P(\exp(sX) \geq \exp(s\epsilon)) \leq e^{-s\epsilon}\mathbb{E}[e^{sX}]$$

for $s > 0$. To prove Proposition 46, we use the following lemma:

Lemma 7 (Hoeffding) *Suppose that a random variable X satisfies $\mathbb{E}[X] = 0$ for $a \leq X \leq b$. Then, for an arbitrary $\epsilon > 0$, we have*

$$\mathbb{E}\left[e^{\epsilon X}\right] \leq e^{\epsilon^2 (b-a)^2/8} \tag{4.32}$$

Proof: See the Appendix at the end of this chapter.

Returning to the proof of Proposition 46, let $S_n := \sum_{i=1}^{n} X_i$, and apply Lemma 5 to obtain

$$P(S_n - \mathbb{E}[S_n] \geq \epsilon) \leq \min_{s>0} e^{-s\epsilon}\mathbb{E}[\exp\{s(S_n - \mathbb{E}[S_n])\}].$$

In particular, since $X_1, \ldots, X_n$ are independent, we have

$$e^{-s\epsilon}\mathbb{E}[\exp\{s(S_n - \mathbb{E}[S_n])\}] = e^{-s\epsilon} \prod_{i=1}^{n} \mathbb{E}[e^{s(X_i - \mathbb{E}[X_i])}].$$

Moreover, by applying Lemma 7, we obtain

$$P(S_n - \mathbb{E}[S_n] \geq \epsilon) \leq \min_{s>0} \exp\{-s\epsilon + \frac{s^2}{8} \sum_{i=1}^{n} (b_i - a_i)^2\}$$

in which the minimum value is attained when $s := 4\epsilon / \sum_{i=1}^{n} (b_i - a_i)^2$, and we have

$$P(S_n - \mathbb{E}[S_n] \geq \epsilon) \leq \exp\{-2\epsilon^2 / \sum_{i=1}^{n} (b_i - a_i)^2\}.$$

Furthermore, if we replace $X_1, \ldots, X_n$ with $-X_1, \ldots, -X_n$, we obtain

$$P(S_n - \mathbb{E}[S_n] \leq -\epsilon) \leq \exp\{-2\epsilon^2 / \sum_{i=1}^{n} (b_i - a_i)^2\}.$$

Hence, we have

$$P(|S_n - \mathbb{E}[S_n]| \geq \epsilon) = 1 - P(|S_n - \mathbb{E}[S_n]| \leq \epsilon)$$

$$\leq P(S_n - \mathbb{E}[S_n] \geq \epsilon) + P(S_n - \mathbb{E}[S_n] \leq -\epsilon) \leq 2\exp\{-2\epsilon^2 / \sum_{i=1}^{n} (b_i - a_i)^2\}.$$

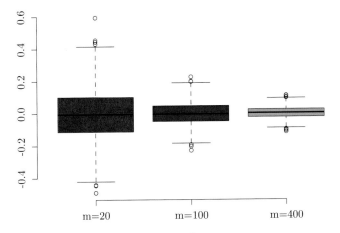

Fig. 4.6 In the RFF approximation, we generated $\hat{k}(x, y)$ 1000 times by changing m. We observe that they all have zero centers, and the larger m is, the smaller the estimation error is

If we substitute $\bar{X} = S_n/n$, we obtain Proposition 46. ☐

Since $\mathbb{E}[\hat{k}(x, y)] = k(x, y)$ and $-2 \le z_i(x)z_i(y) \le 2$, using Proposition 46, we obtain (4.29)[4].

Example 68 From Example 19, since the probability of a Gaussian kernel has a mean of 0 and a covariance matrix $\sigma^{-2}I \in \mathbb{R}^{d \times d}$, we generate the d-dimensional random numbers and uniform random numbers independently and construct the m functions $z_i(x) = \sqrt{2}\cos(\omega_i^\top x + b_i), i = 1, \ldots, m$. We draw a boxplot of $\hat{k}(x, y) - k(x, y)$ by generating (x, y) 1000 times with $d = 1$ and $m = 20, 100, 400$ in Fig. 4.6. We observe that $\hat{k}(x, y) - k(x, y)$ has a mean of 0 ($\hat{k}(x, y)$ is an unbiased estimator), and the larger m is, the smaller the variance is. The program is written as follows.

```
sigma=10; sigma2=sigma^2
k=function(x,y) exp(-(x-y)^2/2/sigma2)
z=function(x) sqrt(2/m)*cos(w*x+b)
zz=function(x,y) sum(z(x)*z(y))
u=matrix(0,1000,3)
m_seq=c(20,100,400)
for(i in 1:1000){
  x=rnorm(1); y=rnorm(1)
  for(j in 1:3){
    m=m_seq[j]; w=rnorm(m)/sigma; b=runif(m)*2*pi
    u[i,j]=zz(x,y)-k(x,y)
  }
}
```

---

[4] The original paper by Rahimi and Recht [23] and subsequent work proved more rigorous upper and lower bounds than these [2].

```
14  boxplot(u[,1],u[,2],u[,3], ylab="the difference from k(x,y)",names=
        paste0("m=",m_seq),
15          col=c("red","blue","green"), main="Kernel Approximation via
        RFF")
```

The solution $\alpha = [\alpha_1, \ldots, \alpha_N]$ with $f(\cdot) = \sum_{i=1}^{N} \alpha_i k(x_i, \cdot)$ for kernel ridge regression with the Gram matrix K is given by (4.6) (Section 4.1). If we obtain the $\hat{f}$ that approximates f and approximates the Gram matrix K via RFF as $\hat{K} = ZZ^\top$, then we obtain $\hat{f}(\cdot) = \sum_{i=1}^{N} \hat{\alpha}_i \hat{k}(x_i, \cdot)$ by using $\hat{\alpha} \in \mathbb{R}^N$ for $(\hat{K} + \lambda I_N)\hat{\alpha} = y$ for $Z = (Z_j(x_i)) \in \mathbb{R}^{N \times m}$ and the unit $I_N \in \mathbb{R}^{N \times N}$.

Using Woodbury's formula, for $U \in \mathbb{R}^{r \times s}$, $V \in \mathbb{R}^{s \times r}$, $r, s \geq 1$,

$$U(I_s + VU) = (I_r + UV)U.$$

And we have

$$Z^\top(ZZ^\top + \lambda I_N)^{-1} = (Z^\top Z + \lambda I_m)^{-1} Z^\top.$$

Let $x \in E$ be a value other than the $x_1, \ldots, x_N$ used for estimation, and let $z(x) := [z_1(x), \ldots, z_m(x)]$ (row vector). Then, for

$$\hat{\beta} := (Z^\top Z + \lambda I_m)^{-1} Z^\top y \tag{4.33}$$

we have

$$\hat{f}(x) = \sum_{i=1}^{N} \alpha_i \hat{k}(x, x_i) = z(x) \sum_{i=1}^{N} z^\top(x_i)\hat{\alpha}_i = z(x)Z^\top\hat{\alpha} = z(x)Z^\top(\hat{K} + \lambda I_N)^{-1} y$$
$$= z(x)(Z^\top Z + \lambda I_m)^{-1} Z^\top y = z(x)\hat{\beta}.$$

Then, for the new $x \in E$, we can find its value from $\hat{f}(x) = z(x)\hat{\beta}$. The computational complexity of (4.33) is $O(m^2 N)$ for the multiplication of $Z^\top Z$, $O(m^3)$ for finding the inverse of $Z^\top Z + \lambda I_m \in \mathbb{R}^{m \times m}$, $O(Nm)$ for the multiplication of $Z^\top y$, and $O(m^2)$ for multiplying $(Z^\top Z + \lambda I_m)^{-1}$ and $Z^\top y$. Thus, overall, the process requires only $O(N^2 m)$ complexity at most. On the other hand, the process takes $O(N^3)$ time when using the kernel without approximation. If $m = N/10$, the computational time becomes 1/100. Obtaining $\hat{f}(x)$ from a new $x \in E$ also takes only $O(m)$ time.

Example 69 We applied RFF to kernel Ridge regression. For $N = 200$ data, we used $m = 20$ for the approximation. We plotted the curve for $\lambda = 10^{-6}$, 10^{-4} (Fig. 4.7). The program is as follows.

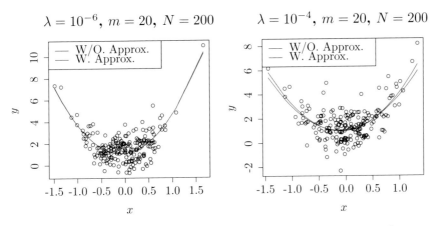

Fig. 4.7 We applied RFF to kernel ridge regression. On the left and right are $\lambda = 10^{-6}$ and $\lambda = 10^{-4}$, respectively

```
sigma=10; sigma2=sigma^2

## Function z
m=20; w=rnorm(m)/sigma; b=runif(m)*2*pi
z=function(u,m) sqrt(2/m)*cos(w*u+b)

## Gaussian kernel
k.g=function(x,y)exp(-(x-y)^2/2/sigma2)

## Data Generation
n=200; x=rnorm(n)/2; y=1+5*sin(x/10)+5*x^2+rnorm(n)
x.min=min(x); x.max=max(x); y.min=min(y); y.max=max(y)
lambda=0.001   ## lambda=0.9

# Low Rank Approximation Function
alpha.rff=function(x,y,m){
  n=length(x)
  Z=array(dim=c(n,m))
  for(i in 1:n)Z[i,]=z(x[i],m)
  beta=solve(t(Z)%*%Z+lambda*diag(m))%*%t(Z)%*%y
  return(as.vector(beta))
}
# Usual Function
alpha=function(k,x,y){
  n=length(x); K=matrix(0,n,n); for(i in 1:n)for(j in 1:n)K[i,j]=k(x
    [i],x[j])
  alpha=solve(K+lambda*diag(n))%*%y
  return(as.vector(alpha))
}
# Numerically compare
alpha.hat=alpha(k.g,x,y)
beta.hat=alpha.rff(x,y,m)
```

```
32  r=sort(x); u=array(n); v=array(n)
33  for(j in 1:n){
34    S=0;for(i in 1:n)S=S+alpha.hat[i]*k.g(x[i],r[j]); u[j]=S
35    v[j]=sum(beta.hat*z(r[j],m))
36  }
37  plot(r,u,type="l",xlim=c(x.min,x.max),ylim=c(y.min,y.max),xlab="x",
        ylab="y",col="red",
38        main="lambda=10^{-4},m=20,n=200")
39  lines(r,v,col="blue"); points(x,y)
40  legend("topleft",lwd=1,c("w/o Approx","w Approx"), col=c("red","blue
        "))
```

The RFF are said to have no significant degradation due to approximation in practice. Still, this does cause an issue regarding theoretical guarantees.

4.6 Nyström Approximation

We consider finding the coefficient estimates $(K + \lambda I)^{-1} y$ in kernel ridge regression. Suppose that we can realize a low-rank matrix decomposition of $K = RR^\top$ with $R \in \mathbb{R}^{N \times m}$ in a computationally inexpensive way. In this case, we can complete the estimation task quickly. Note that we have

$$(RR^\top + \lambda I_N)^{-1} = \frac{1}{\lambda}\{I_N - R(R^\top R + \lambda I_m)^{-1} R^\top\} \tag{4.34}$$

which is due to Sherman-Morrison-Woodbury's formula[5]: $r, s \geq 1$, $A \in \mathbb{R}^{s \times s}$, $U \in \mathbb{R}^{s \times r}$, $C \in \mathbb{R}^{r \times r}$, $V \in \mathbb{R}^{r \times s}$

$$(A + UCV)^{-1} = A^{-1} - A^{-1}U(C^{-1} + VA^{-1}U)^{-1}VA^{-1} \tag{4.35}$$

with $r = m$, $s = N$, $A = \lambda I_N$, $U = R$, $C = I_r$, and $V = R^\top$.

Computing the left side of (4.34) requires an inverse matrix operation of size N, while computing the right side involves the product of $N \times m$ and $m \times m$ matrices and an inverse matrix operation of size m. The computations on the left- and right-hand sides require $O(N^3)$ and $O(N^2 m)$ complexity, respectively. In the following part of this section, we show that with some approximation, the decomposition of $K = RR^\top$ is completed in $O(Nm^2)$ time, i.e., the calculation of the ridge regression is performed in $O(Nm^2)$. In other words, if $N/m = 10$, the computational time is only 1/100.

In Section 3.3, based on (3.18), we considered approximating the eigenfunctions from $x_1, \ldots, x_m \in E$ by

---

[5] Joe Suzuki, "Statistical Learning with Math and R/Python".

$$\phi_i(\cdot) = \frac{\sqrt{m}}{\lambda_i^{(m)}} \sum_{j=1}^{m} k(x_j, \cdot) U_{j,i}.$$

Let $m \leq N$; from the first m samples

$$x_1, \dots, x_m$$

of $x_1, \dots, x_m, x_{m+1}, \dots, x_N$, we construct ϕ_i and λ_i. Then, via

$$v_i := [\phi_i(x_1)/\sqrt{N}, \dots, \phi_i(x_N)/\sqrt{N}] \in \mathbb{R}^N$$

$$\lambda_i^{(N)} := N\lambda_i$$

$$K_N = \sum_{i=1}^{m} \lambda_i^{(N)} v_i v_i^\top$$

we approximate the Gram matrix K_N w.r.t. $x_1, \dots, x_N$. In order to decompose $RR^\top$, we may set it as

$$R = \sqrt{\lambda_i^{(N)}} [v_1, \dots, v_m].$$

To compute R, we require $O(m^3)$ and $O(Nm^2)$ time complexities for obtaining the eigenvalue and eigenvector of K_m and $v_1, \dots, v_m \in \mathbb{R}^N$, respectively. Thus, the computation completes $O(Nm^2)$ time in total..

Example 70 We compared the results of kernel ridge regression with $N = 300$, $m = 10, 20$ and $\lambda = 10^{-5}, 10^{-3}$ (Fig. 4.8). For these data, when $\lambda \geq 1$, the graphs obtained with and without approximation were consistent. For $m = 10, 20$, the curves were almost identical. We observed that the approximation error was smaller when λ was small for RFF, while the error was smaller when λ was large for the Nyström approximation.

```
1  sigma2=1; k.g=function(x,y)exp(-(x-y)^2/2/sigma2)
2  n=300; x=rnorm(n)/2; y=3-2*x^2+3*x^3+2*rnorm(n)        ## Data
      Generation
3  lambda=10^(-5)   ## lambda=0.9
4  m=10
5  # Low rank approximated function
6  alpha.m=function(k,x,y,m){
7    n=length(x); K=matrix(0,n,n); for(i in 1:n)for(j in 1:n)K[i,j]=k(x
         [i],x[j])
8    A=svd(K[1:m,1:m])
9    u=array(dim=c(n,m));
10   for(i in 1:m)for(j in 1:n)u[j,i]=sqrt(m/n)*sum(K[j,1:m]*A$u[1:m,i
         ])/A$d[i]
11   mu=A$d*n/m
12   R=sqrt(mu[1])*u[,1]; for(i in 2:m)R=cbind(R,sqrt(mu[i])*u[,i])
```

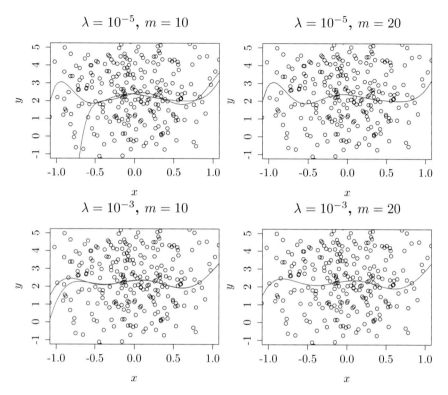

Fig. 4.8 We approximated data with $N = 300$ and ranks $m = 10, 20$. The upper and lower subfigures display the results obtained when running $\lambda = 10^{-5}$ and $\lambda = 10^{-3}$, respectively. The red and blue lines are the results obtained without approximation and with approximation, respectively. The accuracy is almost the same as that in the case without approximation when $m = 20$. The larger the value of λ is, the smaller the approximation error becomes

```
13    alpha=(diag(n)-R%*%solve(t(R)%*%R+lambda*diag(m))%*%t(R))%*%y/
         lambda
14    return(as.vector(alpha))
15  }
16  # Usual function
17  alpha=function(k,x,y){
18    n=length(x); K=matrix(0,n,n); for(i in 1:n)for(j in 1:n)K[i,j]=k(x
         [i],x[j])
19    alpha=solve(K+lambda*diag(n))%*%y
20    return(as.vector(alpha))
21  }
22  # Numerically compare
23  alpha.1=alpha(k.g,x,y); alpha.2=alpha.m(k.g,x,y,m)
24  z=sort(x); u=array(n); v=array(n)
25  for(j in 1:n){
26    S=0;for(i in 1:n)S=S+alpha.1[i]*k.g(x[i],z[j]); u[j]=S
27    S=0;for(i in 1:n)S=S+alpha.2[i]*k.g(x[i],z[j]); v[j]=S
```

```
28  }
29  plot(z,u,type="l",xlim=c(-1,1),xlab="x", ylab="y", ylim=c(-1,5),col
       ="red",main="Kernel Ridge")
30  lines(z,v,col="blue"); points(x,y)
31  legend("topleft",lwd=1,c("w/o Approx","w Approx"), col=c("red","blue
       "))
```

4.7 Incomplete Cholesky Decomposition

In general, we can decompose a positive definite matrix $A \in \mathbb{R}^{N \times N}$ into $A = RR^{\top}$ By using a lower triangular matrix R with nonnegative diagonal components. Such a decomposition is called the Cholesky decomposition of A.

Proposition 47 *For a positive definite matrix $A \in \mathbb{R}^{n \times n}$, there exists a Cholesky decomposition $A = RR^{\top}$ that is unique if and only if A is positive definite.*

Many books cover this material. For the proofs, see, for example, [9].

The following is the Cholesky decomposition procedure. We construct the process so that we can stop anytime to obtain an approximation of $RR^{\top}$ with rank $r \leq N$.

1. In the initial stage, $B = A$, and R is a zero matrix.
2. For each $i = 1, \ldots, r$, the first i columns of R are set so that $B_{j,i} = \sum_{h=1}^{N} R_{j,h} R_{i,h}$ for $j = 1, \ldots, N$. In other words, the setup is complete through the i−th column of B.

$$
R = \begin{bmatrix}
R_{1,1} & 0 & \cdots & & \cdots & 0 \\
\vdots & \ddots & \ddots & & \cdots & 0 \\
R_{i,1} & \vdots & R_{i,i} & 0 & \cdots & 0 \\
R_{i+1,1} & \vdots & R_{i+1,i} & 0 & \cdots & 0 \\
\vdots & \vdots & \vdots & \vdots & \ddots & \vdots \\
R_{N,1} & \cdots & R_{N,i} & 0 & \cdots & 0
\end{bmatrix}
$$

In this case, we swap the two subscripts in B by multiplying a matrix Q from the front and rear of B.

3. The final result is that $RR^{\top} = B = P^{\top}AP$ with $P = Q_1 \cdots Q_N$. Therefore, $A = PRR^{\top}P^{\top}$, and we have that $PR(PR)^{\top}$ is the Cholesky decomposition.

Here, to replace the (i, j) rows and (i, j) columns of the symmetric matrix B, let Q be the matrix obtained by replacing the (i, j), (j, i) and (i, i), (j, j) components of the unit matrix with 1 and 0, respectively, and multiplying B by the symmetric matrix Q from the front and rear of B. For example,

$$
QBQ = \begin{bmatrix} 1 & 0 & 0 \\ 0 & 0 & 1 \\ 0 & 1 & 0 \end{bmatrix} \begin{bmatrix} b_{11} & b_{12} & b_{13} \\ b_{21} & b_{22} & b_{23} \\ b_{31} & b_{32} & b_{33} \end{bmatrix} \begin{bmatrix} 1 & 0 & 0 \\ 0 & 0 & 1 \\ 0 & 1 & 0 \end{bmatrix} = \begin{bmatrix} b_{11} & b_{13} & b_{12} \\ b_{31} & b_{22} & b_{32} \\ b_{21} & b_{23} & b_{33} \end{bmatrix} .
$$

Specifically, for $i = 1, 2, \cdots, r$, we perform the following steps. Assume that $\epsilon > 0$.

1. Let k be the j $(i \le j \le N)$ that maximizes $R_{j,j}^2 = B_{j,j} - \sum_{h=1}^{i-1} R_{j,h}^2$.

   (a) Swap the $i-$th and $k-$th rows and $i-$th and $k-$th columns of B.
   (b) Let $Q_{i,k} := 1$, $Q_{k,i} := 1$, $Q_{i,i} := 0$, $Q_{k,k} := 0$.
   (c) Swap $R_{i,1}, \cdots, R_{i,i-1}$ and $R_{k,1}, \cdots, R_{k,i-1}$.
   (d) $R_{i,i} = \sqrt{B_{k,k} - \sum_{h=1}^{i-1} R_{k,h}^2}$.

2. End if $R_{i,i} < \epsilon$.

3. $R_{j,i} = \dfrac{1}{R_{i,i}}(B_{j,i} - \sum_{h=1}^{i-1} R_{j,h} R_{i,h})$ for each $j = i+1, \cdots, N$.

Once the $i-$th column is completed, $B_{j,i} = \sum_{h=1}^{N} R_{j,h} R_{i,h}$ follows, and $R_{j,i}$ remains the same after that for each $j = 1, \ldots, N$. Then, $B = RR^\top$ follows if the procedure completes up to $r = N$.

At the beginning of each $i = 1, 2, \ldots, r$, we select the j that maximizes $R_{j,j}^2 = B_{j,j} - \sum_{h=1}^{i-1} R_{j,h}^2 \ge 0$. In step 3, the components of the $j-$th $(j = i+1, \ldots, N)$ rows of the $i-$th column join, but we divide them by $R_{i,i}$. Compared to the case where other values are selected as $R_{i,i}$ in step 1, the absolute value of $R_{j,i}$ after dividing by R_{ii} becomes smaller, and the $B_{j,j} - \sum_{h=1}^{i} R_{j,h}^2$ in the next step becomes larger for each j. If $R_{r,r}^2$ takes a negative value, then regardless of the selection order, there is no solution to the Cholesky decomposition, contradicting Proposition 47 (the uniqueness of the solution is also guaranteed). Even in the case of an incomplete Cholesky decomposition, we use the first r columns when running $r = N$.

We show the code for executing the incomplete Cholesky decomposition below.

```
im.ch=function(A,m=ncol(A)){
    n=ncol(A); R=matrix(0,n,n); P=diag(n)
    for(i in 1:n)R[i,i]=sqrt(A[i,i])
    max.R=0;for(i in 1:n)if(R[i,i]>max.R){k=i; max.R=R[i,i]}
    R[1,1]=max.R
    if(k != 1){
        w=A[,k]; A[,k]=A[,1]; A[,1]=w
        w=A[k,]; A[k,]=A[1,]; A[1,]=w
        P[1,1]=0; P[k,k]=0; P[1,k]=1; P[k,1]=1
    }
    for(i in 2:n)R[i,1]=A[i,1]/R[1,1]
    if(m>1)for(i in 2:m){
        for(j in i:n)R[j,j]=sqrt(A[j,j]-sum(R[j,1:(i-1)]^2))
        max.R=0;for(j in i:n)if(R[j,j]>max.R){k=j; max.R=R[j,j]}
        R[i,i]=max.R
        if(k!=i){
            w=R[i,1:(i-1)]; R[i,1:(i-1)]=R[k,1:(i-1)]; R[k,1:(i-1)
]=w
            w=A[,k]; A[,k]=A[,i]; A[,i]=w
            w=A[k,]; A[k,]=A[i,]; A[i,]=w
```

```
20          Q=diag(n); Q[i,i]=0; Q[k,k]=0; Q[i,k]=1; Q[k
     ,i]=1; P=P%*%Q
21        }
22        if(i<n)for(j in (i+1):n)R[j,i]=(A[j,i]-sum(R[i,1:(i-1)]*R[j
     ,1:(i-1)]))/R[i,i]
23      }
24      if(m<n)for(i in (m+1):n)R[,i]=0
25      return(list(P=P,R=R))
26    }
27
28    # Data Generation
29    n=4; range=-5:5
30    D=matrix(sample(range,n*n,replace=TRUE),n,n)
31    A=t(D)%*%D
32
33    # Execution
34    im.ch(A)
35    L=im.ch(A)$R; L%*%t(L)
36    P=im.ch(A)$P; t(P)%*%A%*%P
37    P%*%(L%*%t(L))%*%t(P)    ## Coincides with A
38
39    im.ch(A,2)
40    L=im.ch(A,2)$R; L%*%t(L)
41    P=im.ch(A,2)$P; t(P)%*%A%*%P
42    P%*%(L%*%t(L))%*%t(P)     ## Low Rank Appoximation of A
```

Appendix

Proof of Proposition 44

Since r is a natural spline function whose highest order is $2q - 1$ and it satisfies

$$r^{(q)}(0) = \cdots = r^{(2q-1)}(0) = r^{(q)}(1) = \cdots = r^{(2q-1)}(1) = 0$$

we have

$$\int_0^1 r^{(q)}(x)s^{(q)}(x)dx = \left[r^{(q)}(x)s^{(q-1)}(x)\right]_0^1 - \int_0^1 r^{(q+1)}(x)s^{(q-1)}(x)dx$$

$$= -\int_0^1 r^{(q+1)}(x)s^{(q-1)}(x)dx = \cdots = (-1)^{q-1}\int_0^1 r^{(2q-1)}(x)s'(x)dx$$

$$= (-1)^{q-1}\sum_{j=1}^{N-1} r^{(2q-1)}(x_j^+)\left[s(x_{j+1}) - s(x_j)\right] = 0 \tag{4.36}$$

where we use $s(x_i) = 0$ for $i = 1, \ldots, N$. Moreover, $r^{(2q-1)}(x_j^+)$ is the $(2q - 1)$-th right differential coefficient of r, and it has a constant value during $x_j < x < x_{j+1}$.

Thus, we have the following inequality in the proposition:

$$\int_0^1 \{g^{(q)}(x)\}^2 dx = \int_0^1 \{r^{(q)}(x) + s^{(q)}(x)\}^2 dx$$

$$= \int_0^1 \{r^{(q)}(x)\}^2 dx + \int_0^1 \{s^{(q)}(x)\}^2 dx + 2\int_0^1 r^{(q)}(x)s^{(q)}(x)dx$$

$$= \int_0^1 \{r^{(q)}(x)\}^2 dx + \int_0^1 \{s^{(q)}(x)\}^2 dx \geq \int_0^1 \{r^{(q)}(x)\}^2 dx \qquad (4.37)$$

where the third equality is due to (4.36). On the other hand, from $g, r \in W_q[0, 1]$ and $s \in W_q[0, 1]$, we have

$$s(x) = \sum_{i=0}^{q-1} \frac{s^{(i)}(0)}{i!} x^i + \int_0^1 \frac{(x-u)_+^{q-1}}{(q-1)!} s^{(q)}(u)du.$$

Therefore, when the equality of (4.37) holds, i.e., $\int_0^1 \{s^{(q)}(x)\}^2 dx = 0$, we have $s^{(q)}(x) = 0$ almost everywhere. Hence,

$$s(x) = \sum_{i=0}^{q-1} \frac{s^{(i)}(0)}{i!} x^i$$

which means that $s(x_i) = 0$ for $i = 1, 2, \ldots, N$. Thus, if N exceeds the order of the polynomial $q - 1$, then we require $s(x) = 0$ for $x \in [0, 1]$. □

Proof of Proposition 45

From the additive theorem, we have

$$2\cos(\omega^\top x + b)\cos(\omega^\top y + b) = \cos(w^\top(x - y)) + \cos(w^\top(x + y) + 2b).$$

Since the expectation of the second term w.r.t. b when fixing ω is zero, we have

$$\mathbb{E}_{\omega,b}[\sqrt{2}\cos(\omega^\top x + b) \cdot \sqrt{2}\cos(\omega^\top y + b)] = \mathbb{E}_\omega \cos(w^\top(x - y)).$$

If we apply Euler's formula $e^{i\theta} = \cos\theta + i\sin\theta$ to Proposition 5, then $k(x, y)$ takes a real value. Thus, we have $\mathbb{E}[\sin(\omega^\top(x - y))] = 0$, and $k(x, y)$ can be written as

$$\mathbb{E}_\omega \exp(i\omega^\top(x - y)) = \mathbb{E}_\omega[\cos(\omega^\top(x - y)) + i\sin(\omega^\top(x - y))] = \mathbb{E}_\omega[\cos(\omega^\top(x - y))].$$

From Proposition 5, we obtain (4.28). □

Proof of Lemma 7

Let $\epsilon > 0$. Since $e^{\epsilon x}$ is convex w.r.t. x, if we take the expectation on the both sides of

$$e^{\epsilon X} \leq \frac{X - a}{b - a} e^{\epsilon b} + \frac{b - X}{b - a} e^{\epsilon a}$$

for $b > a$, then

$$\mathbb{E}[e^{\epsilon X}] \leq \frac{-a}{b - a} e^{\epsilon b} + \frac{b}{b - a} e^{\epsilon a} = \theta e^{\epsilon(1-\theta)(b-a)} + (1 - \theta) e^{-\epsilon\theta(b-a)} = \exp\{-\theta s + \log(1 - \theta + \theta e^s)\}$$

for $s = \epsilon(b - a)$ and $\theta = \dfrac{-a}{b - a}$. Therefore, it is sufficient for the exponent $f(s) := -\theta s + \log(1 - \theta + \theta e^s)$ to be at most $s^2/8$. Since

$$f'(s) = -\theta + \frac{\theta e^s}{1 - \theta + \theta e^s}$$

and $f(0) = f'(0) = 0$, we have

$$f''(s) = \frac{(1 - \theta) \cdot \theta e^s}{(1 - \theta + \theta e^s)^2} = \phi(1 - \phi) \leq \frac{1}{4}$$

for $\phi = \dfrac{\theta e^s}{1 - \theta + \theta e^s}$. Hence, a $\mu \in \mathbb{R}$ exists such that

$$f(s) = f(0) + f'(0)(s - 0) + \frac{1}{2}f''(\mu)(s - 0)^2 \leq \frac{s^2}{8}$$

which implies (4.31). □

Exercises 46~64

46 Let k be a kernel and $(x_1, y_1), \ldots, (x_N, y_N)$ be samples, and let $f(\cdot) := \sum_{i=1}^{N} \alpha_i k(x_i, \cdot)$. If we minimize $\sum_{i=1}^{N} \{y_i - f(x_i)\}^2 + \lambda \|f\|^2$, $\lambda > 0$ (kernel ridge regression),why does this mean that we have minimized over $f \in H$? In addition, express the optimal value of $\alpha = [\alpha_1, \ldots, \alpha_N]^\top$ using the Gram matrix $K \in \mathbb{R}^{N \times N}$ and $y = [y_1, \ldots, y_N]^\top$.

47 In kernel PCA, let k be a kernel and $x_1, \ldots, x_N$ be samples, and let $f(\cdot) := \sum_{i=1}^{N} \alpha_i k(x_i, \cdot)$. If we maximize (4.8), why does this mean that we have maximized it over $f \in H$? Additionally, express the eigenequations obtained when $\beta = K^{1/2}\alpha$ by using the Gram matrix $K \in \mathbb{R}^{N \times N}$.

48 In kernel PCA, we wish to find α for a centered Gram matrix, as in (4.9). Complete the function kernel.pca.train by filling in the space below.

```
1   kernel.pca.train=function(x,k){
2   ##  Obtain the Gram matrix from data x and kernel k
3     res=eigen(K)
4     alpha=matrix(0,n,n)
5     for (i in 1:n)alpha[,i]=res$vector[,i]/res$value[i]^0.5
6     return(alpha)
7   }
```

Based on the α obtained from the data X, kernel k, and function `kernel.pca.train`, we wish to calculate the score of $z \in \mathbb{R}^{N \times p}$ (any of the $x_1 \ldots, x_N$) (up to $1 \leq m \leq p$ dimensions). Complete the function below.

```
1   kernel.pca.test=function(x,k,alpha,m,z){
2   ##  Obtain the m−th order score from x,k,alpha,m,z
3     return(pca)
4   }
```

Check whether the constructed function works with the following program.

```
1   sigma.2=0.01; k=function(x,y) exp(−norm(x−y,"2")^2/2/sigma.2)
2   x=as.matrix(USArrests); n=nrow(x); p=ncol(x)
3   alpha=kernel.pca.train(x,k)
4   z=array(dim=c(n,2)); for(i in 1:n)z[i,]=kernel.pca.test(x,k,
        alpha,2,x[i,])
5   min.1=min(z[,1]); min.2=min(z[,2]); max.1=max(z[,1]); max.2=max(
        z[,2])
6   plot(0, xlim=c(min.1,max.1),ylim=c(min.2,max.2),xlab="First",
        ylab="Second",cex.lab=0.75,cex.axis = 0.75, main="Kernel PCA
        (Gauss 0.01)")
7   for(i in 1:n)if(i!=5)text(z[i,1],z[i,2],labels=i,cex = 0.5)
8   text(z[5,1],z[5,2],5,col="red")
```

49   Show that the ordinary PCA and kernel PCA with a linear kernel output the same score.

50   Derive the KKT condition for the kernel SVM (4.12).

51   In Example 66, instead of linear and polynomial kernels, use a Gaussian kernel with different values of σ^2 (three different types), and draw the boundary curve in the same graph.

52   From (4.21) and (4.22), derive $\sum_{j=1}^{J} \beta_{j+4} = 0$ and $\sum_{j=1}^{J} \beta_{j+4}\xi_j = 0$.

53   Prove Proposition 44 according to the following steps.

(a)   Show that $\int_0^1 r^{(q)}(x)s^{(q)}(x)dx = 0$.

(b)   Show that $\int_0^1 \{g^{(q)}(x)\}^2 dx \geq \int_0^1 \{r^{(q)}(x)(x)\}^2 dx$.

(c)   When the equality in (b) holds, show that $s(x) = \sum_{i=0}^{q-1} \frac{s^{(i)}(0)}{i!} x^i$.

(d)   Show that the function s decreases when the equality in (b) holds and N exceeds the degree $q - 1$ of the polynomial.

54 In RFF, instead of finding the kernel $k(x, y)$, we find its unbiased estimator $\hat{k}(x, y)$. Show that the average of $\hat{k}(x, y)$ is $k(x, y)$. Moreover, construct a function that outputs $\hat{k}(x, y)$ from $(x, y) \in E$ for $m = 100$ by using the constants and functions in the program below. Furthermore, compare the result with the value output by the Gaussian kernel and confirm that it is correct.

```
1  sigma=10; sigma2=sigma^2
2  z=function(x) sqrt(2/m)*cos(w*x+b)
3  zz=function(x,y) sum(z(x)*z(y))
```

55 Derive the Chernoff bound.

56 Show that Proposition 46 implies (4.29).

57 The RFF are based on Bochner's theorem (Proposition 5). What relationship exists between them?

58 In RFF, after randomly generating $(w_1, b_1), \ldots, (w_m, b_m)$, we obtain $Z = (z_j(x_i)) \in \mathbb{R}$ for $i = 1, \ldots, N$ and $j = 1, \ldots, m$. If we use $\hat{K} = ZZ^\top$ rather than $K = (k(x_i, x_j)) \in \mathbb{R}^{N \times N}$, show that $\hat{f}(x) = \sum_{i=1}^m \hat{\alpha}_i \hat{k}(x, x_i)$ $(x \in E)$ can be expressed by $\hat{f}(x) = z(x)\hat{\beta}$ using $\hat{\beta}$ in (4.33). Moreover, prove Woodbury's formula:

$$U(I_s + VU) = (I_r + UV)U$$

for $U \in \mathbb{R}^{r \times s}, V \in \mathbb{R}^{s \times r}, r, s \geq 1$.

59 Evaluate the number of computations required to obtain (4.33) for the RFF. In addition, evaluate the computational complexity of finding $\hat{f}(x)$ for the new $x \in E$.

60 To find the coefficient estimates $(K + \lambda I)^{-1}y$ in kernel ridge regression, we wish to decompose the low-rank matrix $K = RR^\top$ with $R \in \mathbb{R}^{N \times m}$. If we can decompose $K = RR^\top$, evaluate the computations on the left- and right-hand sides, where we assume that finding the inverse of the matrix $A \in \mathbb{R}^{n \times n}$ takes $O(n^3)$.

61 We wish to find the coefficient $\hat{\alpha}$ of the kernel ridge regression by using the Nyström approximation. If we use the left-hand side of (4.34) instead of the right-hand side, what changes would be necessary in the following code?

```
1   alpha.m=function(k,x,y,m){
2     n=length(x); K=matrix(0,n,n); for(i in 1:n)for(j in 1:n)K[i,j
      ]=k(x[i],x[j])
3     A=svd(K[1:m,1:m])
4     u=array(dim=c(n,m));
5     for(i in 1:m)for(j in 1:n)u[j,i]=sqrt(m/n)*sum(K[j,1:m]*A$u[1:
      m,i])/A$d[i]
6     mu=A$d*n/m
7     R=sqrt(mu[1])*u[,1]; for(i in 2:m)R=cbind(R,sqrt(mu[i])*u[,i])
8     alpha=(diag(n)-R%*%solve(t(R)%*%R+lambda*diag(m))%*%t(R))%*%y/
      lambda
9     return(as.vector(alpha))
10  }
```

62 In Step 1 of the procedure for the incomplete Cholesky decomposition, each time we choose the j $(i \leq j \leq N)$ that maximizes $R_{j,j}^2 = B_{j,j} - \sum_{h=1}^{i-1} R_{j,h}^2$ as k. Show that $B_{k,k} - \sum_{h=1}^{i-1} R_{k,h}^2$ in Step 1(d) is nonnegative.

63 Show that when the incomplete Cholesky decomposition process is completed up to the r−th column, we have

$$B_{ji} = \sum_{h=1}^{i} R_{jh} R_{jh}$$

for each $i = 1, \ldots, r$ and $j = i + 1, \ldots, N$.

64 Generate a nonnegative definite matrix of size 5×5 and run `im.ch` to perform the incomplete Cholesky decomposition of rank three.

Chapter 5
The MMD and HSIC

In this chapter, we introduce the concept of random variables $X : E \to \mathbb{R}$ in an RKHS and discuss testing problems in RKHSs. In particular, we define a statistic and its null hypothesis for the two-sample problem and the corresponding independence test. We do not know the distribution according to the null hypothesis under a finite sample in either case. Therefore, we introduce a permutation test and a U-statistic with which we construct the process and run the program. Then, we study the notions of characteristics and universal kernels to learn what kernels are valid for such tests. Finally, we learn about empirical processes, which are often used in the mathematical analyses of machine learning and deep learning methods.

5.1 Random Variables in RKHSs

In Chap. 1, we proved that a function $X : E \to \mathbb{R}$ that takes values in $\mathbb{R}$ is measurable if $\{\omega \in E | X(\omega) \in B\}$ for any Borel set B is an event (element) in $\mathcal{F}$, and we call such an X a random variable.

In the following, we say that a kernel k is measurable if the set of (x, y) such that $k(x, y) \in B$ is an event in $E \times E$, and we assume that any kernel k is measurable. Moreover, in this chapter, the expectation $\mathbb{E}[k(X, X)]$ of $k(x, x) \in \mathbb{R}$, $x \in E$, is bounded, which means that both $\mathbb{E}[\sqrt{k(X, X)}] \le \sqrt{\mathbb{E}[k(X, X)]}$ are bounded.

Proposition 48 *Let $k : E \times E \to \mathbb{R}$ be measurable. Then, the map $\Psi : E \ni x \mapsto k(x, \cdot) \in H$ is measurable. Thus, $k(X, \cdot)$ is a random variable in H for any random variable X that takes values in E.*

Proof: See the appendix at the end of this chapter.

Let $X : E \to \mathbb{R}$ be a random variable. The linear functional $T : H \to \mathbb{R}$ with

$$T(f) := \mathbb{E}[f(X)] = \mathbb{E}[\langle f(\cdot), k(X, \cdot) \rangle_H] \le \mathbb{E}[\|f\|_H \sqrt{k(X, X)}] \le \|f\|_H \mathbb{E}[\sqrt{k(X, X)}]$$

© The Author(s), under exclusive license to Springer Nature Singapore Pte Ltd. 2022
J. Suzuki, *Kernel Methods for Machine Learning with Math and R*,
https://doi.org/10.1007/978-981-19-0398-4_5

satisfies $\dfrac{T(f)}{\|f\|_H} \le \mathbb{E}[\sqrt{k(X,X)}] < \infty$. From Proposition 22, there exists an $m_X \in H$ such that

$$\mathbb{E}[f(X)] = \langle f(\cdot), m_X(\cdot) \rangle_H$$

for any $f \in H$. We call such an m_X the expectation of $k(X, \cdot)$, and we write $m_X(\cdot) = \mathbb{E}[k(X, \cdot)]$. Then, we have

$$\mathbb{E}[\langle f(\cdot), k(X, \cdot) \rangle_H] = \langle f(\cdot), \mathbb{E}[k(X, \cdot)] \rangle_H$$

which means that we can change the order of the inner product and expectation operations. Let E_X, E_Y be sets. We define the tensor product H_0 of RKHSs H_X and H_Y consisting of kernels $k_X : E_X \to \mathbb{R}$ and $k_Y : E_Y \to \mathbb{R}$, respectively, by the set of functions $E_X \times E_Y \to \mathbb{R}$, $f(x, y) = \sum_{i=1}^m f_{X,i}(x) f_{Y,i}(y)$, $f_{X,i} \in H_X$, $f_{Y,i} \in H_Y$ for $(x, y) \in E_X \times E_Y$, and we define the inner product and norm by

$$\langle f, g \rangle_{H_0} = \sum_{i=1}^m \sum_{j=1}^n \langle f_{X,i}, g_{X,j} \rangle_{H_X} \langle f_{Y,i}, g_{Y,j} \rangle_{H_Y}$$

and $\|f\|_{H_0}^2 = \langle f, f \rangle_{H_0}$, respectively for $f = \sum_{j=1}^m f_{X,j} f_{Y,j}$, $f_{X,i} \in H_X$, $f_{Y,i} \in H_Y$ and $g = \sum_{j=1}^n g_{X,j} g_{Y,j}$, $g_{X,j} \in H_X$, $g_{Y,j} \in H_Y$. In fact, we have

$$\langle f, g \rangle_{H_0} = \sum_{i=1}^m \sum_{j=1}^n \sum_r \sum_t \alpha_{i,r} \gamma_{j,t} k_X(x_r, x_t) \sum_s \sum_u \beta_{i,s} \delta_{j,u} k_Y(y_s, y_u)$$

$$= \sum_{i=1}^m \sum_r \sum_s \alpha_{i,r} \beta_{i,s} g(x_r, y_s) = \sum_{j=1}^n \sum_t \sum_u \gamma_{j,t} \delta_{j,u} f(x_t, y_u)$$

for $f_{X,i}(\cdot) = \sum_r \alpha_{i,r} k_X(x_r, \cdot)$, $f_{Y,i}(\cdot) = \sum_s \beta_{i,s} k_Y(y_s, \cdot)$, $g_{X,j}(\cdot) = \sum_t \gamma_{j,t} k_X(x_t, \cdot)$, and $g_{Y,j}(\cdot) = \sum_u \delta_{j,u} k_Y(y_u, \cdot)$, which means that the functions do not depend on the expressions of f, g.

If we complete H_0, we can construct a linear space H consisting of the functions $f = \sum_{i=1}^\infty \sum_{j=1}^\infty a_{i,j} e_{X,i} e_{Y,j}$ such that $\|f\|^2 := \sum_{i=1}^\infty \sum_{j=1}^\infty a_{i,j}^2 < \infty$, and the inner product is $\langle f, g \rangle_H = \sum_{i=1}^\infty \sum_{j=1}^\infty a_{i,j} b_{i,j}$, where $g = \sum_{j=1}^\infty b_{i,j} e_{X,i} e_{Y,j}$ ($\sum_{i=1}^\infty \sum_{j=1}^\infty b_{i,j}^2 < \infty$) and $\{e_{X,i}\}$, $\{e_{Y,j}\}$ are orthonormal bases of H_X, H_Y, respectively. Then, H_0 is a dense subspace in H, and H is a Hilbert space. We say that H_0 is the direct product of H_X, H_Y and write $H_X \otimes H_Y$. H is the set of functions f such that $f(x) := \lim_{n \to \infty} f_n(x)$ for any Cauchy sequence $\{f_n\}$ in H_0 and $x \in E$. The claim follows from a similar discussion as that in Steps 1–5 of Proposition 34.

Proposition 49 (Neveu [22]) *The direct product $H_X \otimes H_Y$ of RKHSs H_X, H_Y with reproducing kernels k_X, k_Y is an RKHS with a reproducing kernel $k_X k_Y$.*

Proof: The derivation utilizes the following steps [1].

1. Show that $|g(x, y)| \leq \sqrt{k_X(x, x)} \sqrt{k_Y(y, y)} \|g\|$ for $g \in H_X \otimes H_Y$ and $x \in E_X$, $y \in E_Y$, which means that H is an RKHS due to Proposition 33.
2. Show that $k(x, \cdot, y, \star) := k_X(x, \cdot) k_Y(y, \star) \in H$ when we fix $x \in E_X, y \in E_Y$.
3. Show that $g(x, y) = \langle f(\cdot, \star), k(x, \cdot, y, \star) \rangle_H$.

For details, consult the proof at the end of this chapter. □

Then, we introduce the notion of expectation w.r.t. the variables X, Y. If we assume that $\mathbb{E}[k_X(X, X)]$ and $\mathbb{E}[k_Y(Y, Y)]$ are finite, then $\mathbb{E}_{XY}[k_X(X, \cdot) k_Y(Y, \cdot)]$ is obtained by taking the expectation of $k_X(x, \cdot) k_Y(y, \cdot) \in H_X \otimes H_Y$ w.r.t. XY:

$$\mathbb{E}_{XY}[\|k_X(X, \cdot) k_Y(Y, \cdot)\|_{H_X \otimes H_Y}] = \mathbb{E}_{XY}[\|k_X(X, \cdot)\|_{H_X} \|k_Y(Y, \cdot)\|_{H_Y}]$$
$$= \mathbb{E}_{XY}[\sqrt{k_X(X, X) k_Y(Y, Y)}] \leq \sqrt{\mathbb{E}_X[k_X(X, X)] \mathbb{E}_Y[k_Y(Y, Y)]}.$$

Thus, the left-hand side takes a finite value, and we have

$$\mathbb{E}_{XY}[f(X, Y)] = \mathbb{E}_{XY}[\langle f, k_X(X, \cdot) k_Y(Y, \cdot) \rangle] \leq \|f\|_{H_X \otimes H_Y} \mathbb{E}_{XY}[\|k_X(X, \cdot) k_Y(Y, \cdot)\|_{H_X \otimes H_Y}]$$

for $f \in H_X \otimes H_Y$. From Proposition 22 (Riesz's representation theorem), there exists an $m_{XY} \in H_X \otimes H_Y$ such that

$$\mathbb{E}_{XY}[f(X, Y)] = \langle f, m_{XY} \rangle,$$

and we write

$$m_{XY} := \mathbb{E}_{XY}[k_X(X, \cdot) k_Y(Y, \cdot)]$$

which means that we can change the order of the inner product and expectation operations:

$$\mathbb{E}_{XY}[\langle f, k_X(X, \cdot) k_Y(Y, \cdot) \rangle] = \langle f, \mathbb{E}_{XY}[k_X(X, \cdot) k_Y(Y, \cdot)] \rangle.$$

Moreover, for the m_X, m_Y of X, Y, the expectation $m_X m_Y$ belongs to $H_X \otimes H_Y$, and we have

$$\langle fg, m_X m_Y \rangle_{H_X \otimes H_Y} = \langle f, m_X \rangle_{H_X} \langle g, m_Y \rangle_{H_Y} = \mathbb{E}_X[f(X)] \mathbb{E}_Y[g(Y)]$$

$f \in H_X, g \in H_Y$, which means that we multiply the expectations of X, Y even if they are not independent. Thus, we call

$$m_{XY} - m_X m_Y$$

the covariate of (X, Y) in $H_X \otimes H_Y$, which belongs to $H_X \otimes H_Y$.

Proposition 50 *For each $f \in H_X, g \in H_Y$, there exist $\Sigma_{XY} \in B(H_Y, H_X)$ and $\Sigma_{YX} \in B(H_X, H_Y)$ such that*

$$\langle fg, m_{XY} - m_X m_Y \rangle_{H_X \otimes H_Y} = \langle \Sigma_{YX} f, g \rangle_{H_Y} = \langle f, \Sigma_{XY} g \rangle_{H_X}. \qquad (5.1)$$

Proof: The operators Σ_{YX}, Σ_{XY} are conjugates of each other, and from Proposition 22, if one exists, so does the other. We prove the existence of Σ_{XY}. The linear functional

$$T_g : H_X \ni f \mapsto \langle fg, m_{XY} - m_X m_Y \rangle_{H_X \otimes H_Y} \in \mathbb{R}$$

for an arbitrary $g \in H_Y$ is bounded from

$$\langle fg, m_{XY} - m_X m_Y \rangle_{H_X \otimes H_Y} \leq \|f\|_{H_X} \|g\|_{H_Y} \|m_{XY} - m_X m_Y\|_{H_X \otimes H_Y}$$

and there exists an $h_g \in H_X$ such that $T_g f = \langle f, h_g \rangle_{H_X}$ from Proposition 22. Thus, there exists $\Sigma_{XY} : H_Y \ni g \mapsto h_g \in H_X$ such that

$$\langle fg, m_{XY} - m_X m_Y \rangle_{H_X \otimes H_Y} = \langle f, \Sigma_{XY} g \rangle_{H_X}.$$

The boundness of Σ_{XY} is due to

$$\|\Sigma_{XY} g\|_{H_X} = \|h_g\|_{H_X} = \|T_g\| \leq \|g\|_{H_Y} \|m_{XY} - m_X m_Y\|_{H_X \otimes H_Y}.$$

$\square$

We call Σ_{XY}, Σ_{YX} the mutual covariance operators.

Let H and k be an RKHS and its reproducing kernel, respectively, and let $\mathcal{P}$ be the set of distributions that X follows. Then, we can define the map

$$\mathcal{P} \ni \mu \mapsto \int k(x, \cdot) d\mu(x) \in H$$

which we call the embedding of probabilities in the RKHS. Suppose that the map is injective, i.e., if the expectations $\int k(x, \cdot) d\mu_1(x)$ and $\int k(x, \cdot) d\mu_2(x)$ have the same value, then the probabilities μ_1, μ_2 coincide. We call such a reproducing kernel k of an RKHS H characteristic.

We learn some applications by using characteristic kernels, such as two-sample problems and independence tests, and we consider the associated theory in later sections of this chapter.

5.2   The MMD and Two-Sample Problem

Gretton et al. [11] proposed a statistical testing approach for testing whether two distributions share given independent sequences $x_1, \ldots, x_m \in \mathbb{R}$ and $y_1, \ldots, y_n \in \mathbb{R}$. We write the two distributions as P, Q and regard $P = Q$ as the null hypothesis. Let H and k be an RKHS and its reproducing kernel, respectively; we define $m_P := \mathbb{E}_P[k(X, \cdot)] = \int_E k(x, \cdot) dP(x), m_Q := \mathbb{E}_Q[k(X, \cdot)] = \int_E k(x, \cdot) dQ(x) \in H$.

We note that the random variable $X : E \to \mathbb{R}$ is measurable, and either P or Q is the probability distribution that X follows.

Let $\mathcal{F}$ be a set of functions that satisfies a condition. In general, the quantity defined by

$$\sup_{f \in \mathcal{F}} \{\mathbb{E}_P[f(X)] - \mathbb{E}_Q[f(X)]\}$$

is called the MMD (maximum mean discrepancy), and we assume that

$$\mathcal{F} := \{f \in H \mid \|f\|_H \leq 1\}$$

which means that we regard the MMD as

$$
\begin{aligned}
\mathrm{MMD}^2 &= \sup_{f \in \mathcal{F}} \{\mathbb{E}_P[f(X)] - \mathbb{E}_Q[f(X)]\}^2 = \sup_{f \in \mathcal{F}} \{\langle m_P, f \rangle - \langle m_Q, f \rangle\}^2 \\
&= \sup_{f \in \mathcal{F}} \{\langle m_P - m_Q, f \rangle\}^2 = \|m_P - m_Q\|_H^2 .
\end{aligned}
$$

If the kernel k is characteristic, then we have

$$\mathrm{MMD} = 0 \iff m_P = m_Q \iff P = Q \tag{5.2}$$

and

$$
\begin{aligned}
&\mathrm{MMD}^2 \\
&= \langle m_P, m_P \rangle + \langle m_Q, m_Q \rangle - 2 \langle m_P, m_Q \rangle \\
&= \langle \mathbb{E}_X[k(X, \cdot)], \mathbb{E}_{X'}[k(X', \cdot)] \rangle + \langle \mathbb{E}_Y[k(Y, \cdot)], \mathbb{E}_{Y'}[k(Y', \cdot)] \rangle - 2 \langle \mathbb{E}_X[k(X, \cdot)], \mathbb{E}_Y[k(Y, \cdot)] \rangle \\
&= \mathbb{E}_{XX'}[k(X, X')] + \mathbb{E}_{YY'}[k(Y, Y')] - 2 \mathbb{E}_{XY}[k(X, Y)]
\end{aligned}
$$

where X' and X (Y' and Y) are independent and follow the same distribution. However, we do not know m_X, m_Y from the two-sample data. Thus, we execute the test using their estimates:

$$\widehat{\mathrm{MMD}}_B^2 := \frac{1}{m^2} \sum_{i=1}^m \sum_{j=1}^m k(x_i, x_j) + \frac{1}{n^2} \sum_{i=1}^n \sum_{j=1}^n k(y_i, y_j) - \frac{2}{mn} \sum_{i=1}^m \sum_{j=1}^n k(x_i, y_j)$$

$$\tag{5.3}$$

$$\frac{1}{m(m-1)} \sum_{i=1}^m \sum_{j \neq i} k(x_i, x_j) + \frac{1}{n(n-1)} \sum_{i=1}^n \sum_{j \neq i} k(y_i, y_j) - \frac{2}{mn} \sum_{i=1}^m \sum_{j=1}^n k(x_i, y_j) .$$

$$\tag{5.4}$$

Then, the estimate (5.4) is unbiased, while (5.3) is biased:

$$\mathbb{E}[\frac{1}{m(m-1)} \sum_{i=1}^m \sum_{j \neq i} k(X_i, X_j)] = \frac{1}{m} \sum_{i=1}^m \mathbb{E}_{X_i}[\frac{1}{m-1} \sum_{j \neq i} \mathbb{E}_{X_j}[k(X_i, X_j)]] = \mathbb{E}_{XX'}[k(X, X')] .$$

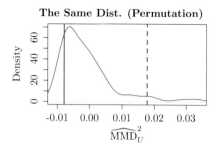
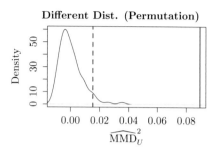

Fig. 5.1 Permutation test for the two-sample problem. The distributions of X, Y are the same (left) and different (right). The blue and red dotted lines show the statistics and the borders of the rejection region, respectively

However, similar to the HSIC in the next section, we do not know the distribution of the MMD estimate under $P = Q$. We consider executing one of the following processes.

1. Construct a histogram of the MMD estimate values randomly by changing the values of $x_1, \ldots, x_m$ and $y_1, \ldots, y_n$ (permutation test).
2. Compute an asymptotic distribution from the distribution of U statistics.

For the former, for example, we may construct the following procedure.

Example 71 We perform a permutation test on two sets of 100 samples that follow the standard Gaussian distribution (Fig. 5.1 Left). For the unbiased estimator of MMD^2, we use $\widehat{MMD}_U^2$ in (5.6) instead of (5.4) for a later comparison. We also double the standard deviation of one set of samples and perform the permutation test again (Fig. 5.1 Right). The reason why $\widehat{MMD}_U^2$ also takes negative values is that when the true value of the MMD is close to zero, the value can also be negative since it is an unbiased estimator.

```
sigma=1; k=function(x,y)exp(-(x-y)^2/sigma^2)
## Data Generation
n=100
xx=rnorm(n)
yy=rnorm(n)           ## The distributions are equal.
# yy=rnorm(n)*2        ## The distributions are not equal.
x=xx;y=yy
## Null Hypothesis
T=NULL
for(h in 1:100){
  index1=sample(n,n/2)
  index2=setdiff(1:n,index1)
  x=c(xx[index2],yy[index1])
  y=c(xx[index1],yy[index2])
  S=0
```

```
16    for(i in 1:n)for(j in 1:n)if(i!=j)S=S+k(x[i],x[j])+k(y[i],y[j])-k
         (x[i],y[j])-k(x[j],y[i])
17    T=c(S/n/(n-1),T)
18    }
19    v=quantile(T,0.95)
20    ## Statistics
21    S=0; for(i in 1:n)for(j in 1:n)if(i!=j)S=S+k(x[i],x[j])+k(y[i],y[j])
         -k(x[i],y[j])-k(x[j],y[i])
22    u=S/n/(n-1)
23    ## Graphical Output
24    plot(density(T),xlim=c(min(T,v,u),max(T,v,u)))
25    abline(v=v,col="red",lty=2,lwd=2)
26    abline(v=u,col="blue",lty=1,lwd=2)
```

For the latter approach, we construct the following quantities. For $m \geq 1$ symmetric variables and $h : E^m \to \mathbb{R}$, we call the quantity

$$U_N := \frac{1}{\binom{N}{m}} \sum_{1 \leq i_1,\ldots,i_m \leq N} h(x_{i_1}, \ldots, x_{i_m}) \tag{5.5}$$

the U statistic w.r.t. h of order m, where $\sum_{i_1,\ldots,i_m}$ ranges over $\binom{N}{m}$ $(i_1,\ldots,i_m) \in \{1,\ldots,N\}^m$'s. We use this quantity for estimating the expectation $\mathbb{E}[h(X_1,\ldots,X_m)]$ given samples $x_1,\ldots,x_N$. Note that any U statistic is unbiased. In fact, we have

$$\mathbb{E}[\frac{1}{\binom{N}{m}} \sum_{i_1<\ldots<i_m} h(X_{i_1},\ldots,X_{i_m})]$$

$$= \frac{1}{\binom{N}{m}} \sum_{i_1<\ldots<i_m} \mathbb{E}h(X_{i_1},\ldots,X_{i_m}) = \mathbb{E}h(X_1,\ldots,X_m).$$

We call the quantity

$$V_N := \frac{1}{N^m} \sum_{i_1=1}^{N} \cdots \sum_{i_m=1}^{N} h(x_{i_1}, \ldots, x_{i_m})$$

the V-statistic w.r.t. h

In the following, we conduct a statistical test with the null hypothesis that X, Y are identically distributed when $m = n$. Under this null hypothesis, the operations of taking the means of $\mathbb{E}_X[\cdot]$ and $\mathbb{E}_Y[\cdot]$ have the same meaning.

We can define an unbiased estimator of MMD^2 in addition to (5.4). In the following, we consider the unbiased estimator

$$\widehat{\text{MMD}}_U^2 = \frac{1}{n(n-1)} \sum_{i \neq j} h(z_i, z_j)$$

for

$$h(z_i, z_j) := k(x_i, x_j) + k(y_i, y_j) - k(x_i, y_j) - k(x_j, y_i) \tag{5.6}$$

with $z_i = (x_i, y_i)$.

We define

$$h_c(z_1, \ldots, z_c) := \mathbb{E}_{Z_{c+1} \cdots Z_m} h(z_1, \ldots, z_c, Z_{c+1}, \ldots, Z_m),$$

which is obtained by taking the expectation of the U statistics (5.5) over $Z_{c+1}, \ldots, Z_m$ for $1 \leq c \leq m$. Moreover, we define

$$\tilde{h}_c(z_1, \ldots, z_c) := h_c(z_1, \ldots, z_c) - \theta$$

for $\theta = \mathbb{E}[h(Z_1, \ldots, Z_m)]$.

Example 72 For (5.6), we have $h_2(z_1, z_2) = h(z_1, z_2)$ since $m = 2$. Under the null hypothesis, X, Y follow the same distribution, and we have

$$h_1(z_1) = \mathbb{E}_{Z_2}[h(z_1, Z_2)] = \mathbb{E}[k(x_i, X_j)] + \mathbb{E}[k(y_i, Y_j)] - \mathbb{E}[k(x_i, Y_j)] - \mathbb{E}[k(x_j, Y_i)] = 0.$$

Moreover, under the null hypothesis, since $\theta = \mathbb{E}h(Z_1, \ldots, Z_m) = 0$, we have $\tilde{h}_2(z_1, z_2) = h(z_1, z_2)$.

Hereafter, we set the number of samples as $N(= m = n)$.

Proposition 51 (Serfling [27]) *Suppose that the U statistics are $\mathbb{E}h^2 < \infty$ and that $h_1(z_1)$ is zero (degenerated). Let $\lambda_1, \lambda_2, \ldots$ be the eigenvalues of the conjugate integral operation*

$$L^2 \ni f(\cdot) \rightarrow \int \hat{h}_2(\cdot, y) f(y) d\eta(y)$$

whose kernel is $\tilde{h}_2(z_1, z_2)$.[1] Then, N times the U statistics converges to the random variable

$$\sum_{j=1}^{\infty} \lambda_j (\chi_j^2 - 1)$$

as $m \rightarrow \infty$, where $\chi_1^2, \chi_2^2, \ldots$ are random variables that are independent of each other and follow a χ^2 distribution with one degree of freedom.

---

[1] The kernel $K : E \times E \rightarrow \mathbb{R}$ in the integral operation $L^2(E, \eta) \ni f \mapsto Kf(\cdot) = \int_E K(\cdot, x) f(x) d\eta(x)$ is called a kernel of the integral operator even if it is not positive definite.

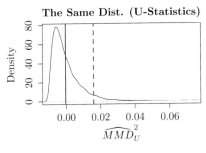 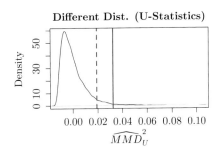

Fig. 5.2 Test performed using the U statistic for the two-sample problem. The same (left) and different (right) distributions of X, Y are employed. The blue line is the statistic, and the red dotted line is the boundary of the rejection region. We can see that the distribution obtained according to the null hypothesis has almost the same shape as that in Fig. 5.1

For the proof, see Sect. 5.5.2 of Serfling [27] (page 193-199).

Note that $\tilde{h}_2(z_1, z_2) = h(z_1, z_2)$ is given by (5.6), which is symmetric but not nonnegative definite. Therefore, Mercer's theorem cannot be applied. However. an integral operator is generally compact (Proposition 39), and if the kernel of an integral operator is symmetric, then the integral operator is self-adjoint (e.g., 45). Therefore, from Proposition 27, eigenvalues and eigenfunctions exist. However, since they are not nonnegative definite, some eigenvalues may not be nonnegative.

In the following, we write $\{\lambda_i\}_{i=1}^{\infty}$ and $\{\phi_i(\cdot)\}_{i=1}^{\infty}$ as the eigenvalues and eigenfunctions, respectively, of the integral operator

$$T_{\tilde{h}} : L^2[E, \mu] \ni f \mapsto \int_E \tilde{h}_2(\cdot, y) f(y) d\eta(y) \in L^2[E, \eta]$$

For the kernel h_2 when $\eta = P = Q$. Then, we have

$$\int_E h_2(x, y)\phi_i(y)d\eta(y) = \lambda_i \phi_i(x)$$

$$\int_E \phi_i(x)\phi_j(x)d\eta(x) = \delta_{i,j}. \tag{5.7}$$

Utilizing Proposition 51, we find that $N\widehat{\text{MMD}}_U^2$ converges to the random variable

$$\sum_{j=1}^{\infty} \lambda_j(\chi_j^2 - 1)$$

as the sample size $N \to \infty$.

Example 73 With two sets of 100 samples that follow the standard Gauss distribution, we obtain eigenvalues by using the method described in Sect. 3.3 and construct a distribution following the null hypothesis using the U statistic to perform the test (Fig. 5.2 Left). We also perform the same test by doubling the standard deviation of one pair of samples (Fig. 5.2 Right).

```
1   sigma=1; k=function(x,y)exp(-(x-y)^2/sigma^2)
2   ## Data Generation
3   n=100
4   x=rnorm(n)
5   y=rnorm(n)           ## The distributions are equal.
6   # y=rnorm(n)*2        ## The distributions are not equal.
7   ## Null Hypothesis
8   K=matrix(0,n,n)
9   for(i in 1:n)for(j in 1:n)K[i,j]=k(x[i],x[j])+k(y[i],y[j])-k(x[i],
10    y[j])-k(x[j],y[i])
11  lambda=eigen(K)$values/n
12  r=20
13  z=NULL
14  for(h in 1:10000)z=c(z,1/n*(sum(lambda[1:r]*(rchisq(1:r, df=1)-1))))
15  v=quantile(z,0.95)
16  ## Statistics
17  S=0
18  for(i in 1:(n-1))for(j in (i+1):n)S=S+k(x[i],x[j])+k(y[i],y[j])-k(x
        [i],y[j])-k(x[j],y[i])
19  u=S/n/(n-1)
20  ## Graphical Output
21  plot(density(z),xlim=c(min(z,v,u),max(z,v,u)))
22  abline(v=v,col="red",lty=2,lwd=2)
23  abline(v=u,col="blue",lty=1,lwd=2)
```

5.3   The HSIC and Independence Test

Let $(E, \mathcal{F}, P)$ be a probability space. We say that events $A, B \in \mathcal{F}$ are independent if $P(A)P(B) = P(A \cap B)$.

Suppose that sequences $x_1, \ldots, x_N \in \mathbb{R}$ and $y_1, \ldots, y_N \in \mathbb{R}$ with the same length $N \geq 1$ have occurred according to the distributions of the random variables X, Y. We wish to test the independence of X, Y, where both x_i, x_j and y_i, y_j are independent, but we do not know whether x_i, y_i are independent.

For example, if the empirical correlation coefficient

$$\hat{\rho} = \frac{(1/N) \sum_{i=1}^{N} (x_i - \bar{x}) \sum_{i=1}^{N} (y_i - \bar{y})}{\{(1/N) \sum_{i=1}^{N} (x_i - \bar{x})^2\}^{1/2} \{(1/N) \sum_{i=1}^{N} (y_i - \bar{y})^2\}^{1/2}}$$

is close to zero for $\bar{x} := (1/N) \sum_{i=1}^{N} x_i$ and $\bar{y} := (1/N) \sum_{i=1}^{N} y_i$, then we may say that the variables are independent.

Example 74 *(Gaussian Distribution)* For simplicity, we assume that X, Y follows the standard Gaussian distribution. If X, Y are independent (written as $X \perp\!\!\!\perp Y$), then their covariance

$$\mathbb{E}[XY] = \int_E \int_E xy f_{XY}(x, y) dx dy = \int_E \int_E xy f_X(x) f_Y(y) dx dy = \int_E x f_X(x) dx \int_E y f_Y(y) dy$$

is 0, and $\rho_{XY} = \mathbb{E}[XY] = 0$ follows. Since we can write $f_{XY}(x, y)$ as

$$\frac{1}{2\sqrt{1 - \rho_{XY}^2}} \exp\{-\frac{1}{2(1 - \rho_{XY}^2)}(x^2 - 2\rho_{XY} xy + y^2)\}$$

we see that $\rho_{XY} = 0$ implies $f_{XY}(x, y) = f_X(x) f_Y(y)$. Hence, $\rho_{XY} = 0 \iff X \perp\!\!\!\perp Y$ follows.

However, as in the following derivation, in general, $\rho_{XY} = 0$ does not mean that $X \perp\!\!\!\perp Y$.

Example 75 Let $X = \cos\theta$ and $Y = \sin\theta$. We uniformly generate random variables $0 \le \theta < 2\pi$, which means that (X, Y) is uniform over the unit circle. If X is determined, then $Y = \pm\sqrt{1 - X^2}$, and if one of X, Y is determined, then at most two possibilities exist for the other. Therefore, the two variables are not independent. However, since the mean of X, Y is $\mu_X = \mu_Y = 0$, the covariance can be calculated as

$$\mathbb{E}_{XY}[(X - \mu_X)(Y - \mu_Y)] = \mathbb{E}_{XY}[XY] = \mathbb{E}_{XY}[\cos\theta \sin\theta] = \frac{1}{2}\mathbb{E}_{XY}[\sin 2\theta] = 0$$

and the correlation coefficient ρ_{XY} is 0.

To this end, Gretton et al. [12] thought that to test the independence of random variables X, Y, if we map $E \ni X \mapsto k_X(X, \cdot) \in H_X$ and $E \ni Y \mapsto k_Y(Y, \cdot) \in H_Y$ for kernels k_X, k_Y, and perform the test of independence based on the covariance between $k_X(X, \cdot)$ and $k_Y(Y, \cdot)$, such an inconvenience would not occur. They devised a statistical test for $\mathbb{E}[k_X(X, \cdot) k_Y(Y, \cdot)] = \mathbb{E}[k_X(X, \cdot)]\mathbb{E}[k_Y(Y, \cdot)]$ rather than $\mathbb{E}[XY] = \mathbb{E}[X]\mathbb{E}[Y]$. We define

$$HSIC(X, Y) := \|m_X m_Y - m_{XY}\|_{H_X \otimes H_Y}^2 \in \mathbb{R}$$

which is the norm of the covariance $m_X m_Y - m_{XY} \in H_1 \otimes H_2$, i.e., the Hilbert-Schmidt information criterion (HSIC). Since the HSIC is a norm, it is zero only if $m_X m_Y - m_{XY} \in H_{X \otimes Y}$ is zero. The HSIC is the MMD^2 when $m_P = m_X m_Y$ with $m_Q = m_{XY}$.

Proposition 52 (Gretton et al. [12]) *When the reproducing kernels $k_X, k_Y : E \to \mathbb{R}$ in H_X, H_Y are both characteristic kernels, the random variables X, Y that take values in E being independent and $HSIC(X, Y) = 0$ are equivalent, i.e.,*

$$HSIC(X, Y) = 0 \Longleftrightarrow X \perp\!\!\!\perp Y.$$

Proof: If k_X, k_Y are both characteristic, then from Proposition 55 (see below), $k_X k_Y$ is also characteristic. Therefore, for $m_X(\cdot) := \int_E k_X(x, \cdot) d P_X(x) \in H_X$. $m_Y(\cdot) := \int_E k_Y(y, \cdot) d P_Y(y) \in H_Y$, and $m_{XY}(\cdot, \star) := \int_E k_X(x, \cdot) k_Y(y, \star) d P_{XY}(x, y) \in H_{X \otimes Y}$, the map $\mathcal{P}_{X \otimes Y} \ni P_{XY} \mapsto m_{XY} \in H_{X \otimes Y}$ is injective. Hence, we have

$$X \perp\!\!\!\perp Y \Longleftrightarrow P_{XY} = P_X P_Y \Longleftrightarrow m_{XY} = m_X m_Y \Longleftrightarrow HSIC(X, Y) = 0$$

where the second $\Longleftrightarrow$ is due to (5.2).   □

From Propositions 50 and 52, we have that

Corollary 2 *both of the reproducing kernels $k_X, k_Y : E \to \mathbb{R}$ of H_X, H_Y are characteristic, and we obtain*

$$\Sigma_{XY} = \Sigma_{YX} = 0 \Longleftrightarrow X \perp\!\!\!\perp Y.$$

If we abbreviate $\| \cdot \|_{X \otimes Y}$ and $\langle \cdot, \cdot \rangle_{X \otimes Y}$ as $\| \cdot \|$ and $\langle \cdot, \cdot \rangle$, respectively, then we have

$$\begin{aligned}
\|m_{XY}\|^2 &= \langle \mathbb{E}_{XY}[k_X(X, \cdot) k_Y(Y, \cdot)], \mathbb{E}_{X'Y'}[k_X(X', \cdot) k_Y(Y', \cdot)] \rangle \\
&= \mathbb{E}_{XY} \mathbb{E}_{X'Y'}[\langle k_X(X, \cdot) k_Y(Y, \cdot), k_X(X', \cdot) k_Y(Y', \cdot) \rangle \\
&= \mathbb{E}_{XYX'Y'}[k_X(X, X') k_Y(Y, Y')]
\end{aligned}$$

$$\begin{aligned}
\langle m_{XY}, m_X m_Y \rangle &= \langle \mathbb{E}_{XY}[k_X(X, \cdot) k_Y(Y, \cdot)], \mathbb{E}_{X'}[k_X(X', \cdot)] \mathbb{E}_{Y'}[k_Y(Y', \cdot)] \rangle \\
&= \mathbb{E}_{XY}\{\mathbb{E}_{X'}[\langle k_X(X, \cdot) k_Y(Y, \cdot), k_X(X', \cdot) \mathbb{E}_{Y'}[k_Y(Y', \cdot)] \rangle]\} \\
&= \mathbb{E}_{XY}\{\mathbb{E}_{X'}[k_X(X, X')] \mathbb{E}_{Y'}[\langle k_Y(Y, \cdot), k_Y(Y', \cdot) \rangle]\} \\
&= \mathbb{E}_{XY}\{\mathbb{E}_{X'}[k_X(X, X')] \mathbb{E}_{Y'}[k_Y(Y, Y')]\}
\end{aligned}$$

and

$$\begin{aligned}
\|m_X m_Y\|^2 &= \langle \mathbb{E}_X[k_X(X, \cdot)] \mathbb{E}_Y[k_Y(Y, \cdot)], \mathbb{E}_{X'}[k_X(X', \cdot)] \mathbb{E}_{Y'}[k_Y(Y', \cdot)] \rangle \\
&= \mathbb{E}_X \mathbb{E}_{X'}[k_X(X, X')] \mathbb{E}_Y \mathbb{E}_{Y'}[k_Y(Y, Y')]
\end{aligned}$$

where X, X' (Y, Y') are independent and follow the same distribution. Hence, we can write $HSIC(X, Y)$ as

$$\begin{aligned}
HSIC(X, Y) &:= \|m_{XY} - m_X m_Y\|^2 \\
&= \mathbb{E}_{XX'YY'}[k_X(X, X') k_Y(Y, Y')] - 2\mathbb{E}_{XY}\{\mathbb{E}_{X'}[k_X(X, X')] \mathbb{E}_{Y'}[k_Y(Y, Y')]\} \\
&\quad + \mathbb{E}_{XX'}[k_X(X, X')] \mathbb{E}_{YY'}[k_Y(Y, Y')].
\end{aligned} \tag{5.8}$$

When applying the HSIC, we often construct the following estimator, replacing the mean by the relative frequency.

$$\widehat{HSIC} := \frac{1}{N^2} \sum_i \sum_j k_X(x_i, x_j) k_Y(y_i, y_j) - \frac{2}{N^3} \sum_i \sum_j k_X(x_i, x_j) \sum_h k_Y(y_i, y_h)$$

$$+ \frac{1}{N^4} \sum_i \sum_j k_X(x_i, x_j) \sum_h \sum_r k_Y(y_h, y_r) \tag{5.9}$$

For example, we can write the HSIC in the R language as follows.

```
HSIC.1=function(x,y,k.x,k.y){
    n=length(x)
    S=0;for(i in 1:n)for(j in 1:n)S=S+k.x(x[i],x[j])*k.y(y[i],y[j])
    T=0;
    for(i in 1:n){
        T.1=0; for(j in 1:n)T.1=T.1+k.x(x[i],x[j]);
        T.2=0; for(l in 1:n)T.2=T.2+k.y(y[i],y[l]);
        T=T+T.1*T.2
    }
    U=0; for(i in 1:n)for(j in 1:n)U=U+k.x(x[i],x[j])
    V=0; for(i in 1:n)for(j in 1:n)V=V+k.y(y[i],y[j])
    return(S/n^2-2*T/n^3+U*V/n^4)
}
```

We often write the statistics as $\widehat{HSIC} = \frac{1}{N^2} \text{trace}(K_X H K_Y H)$, where $K_X = (k_X(x_i, x_j))_{i,j}$, $K_Y = (k_Y(y_i, y_j))_{i,j}$, $H := I - \frac{1}{N} E$, $I \in \mathbb{R}^{N \times N}$ is the unit matrix, and $E \in \mathbb{R}^{N \times N}$ is a matrix in which all the elements are ones. In fact, we have

$$\text{trace}(K_X H K_Y H) = \sum_i (K_X H K_Y H)_{i,i} = \sum_i \sum_j (K_X H)_{i,j}(K_Y H)_{j,i}$$

$$= \sum_i \sum_j \{\sum_h k_X(x_i, x_h)(\delta_{h,j} - \frac{1}{N})\}\{\sum_h k_Y(y_j, y_h)(\delta_{h,i} - \frac{1}{N})\}$$

$$= \sum_i \sum_j \{k_X(x_i, x_j)k_Y(y_i, y_j) - \frac{1}{N}k_X(x_i, x_j)\sum_h k_Y(y_i, y_h)$$

$$- \frac{1}{N}k_Y(y_i, y_j)\sum_h k_X(x_i, x_h) + \frac{1}{N^2}\sum_h k_X(x_i, x_h)\sum_r k_Y(y_j, y_r)\}$$

$$= \sum_i \sum_j k_X(x_i, x_j)k_Y(y_i, y_j) - \frac{2}{N}\sum_i \sum_j k_X(x_i, x_j)\sum_h k_Y(y_i, y_h)$$

$$+ \frac{1}{N^2}\sum_i \sum_h k_X(x_i, x_h)\sum_j \sum_r k_Y(y_j, y_r).$$

```
1  HSIC.1=function(x,y,k.x,k.y){
2      n=length(x)
3      K.x=matrix(0,n,n)
4      for(i in 1:n)for(j in 1:n)K.x[i,j]=k.x(x[i],x[j])
5      K.y=matrix(0,n,n)
6      for(i in 1:n)for(j in 1:n)K.y[i,j]=k.y(y[i],y[j])
7      E=matrix(1,n,n)
8      H=diag(n)−E/n
9      return(sum(diag(K.x%*%H%*%K.y%*%H)))/n^2)
10 }
```

Example 76 We execute the above process for $\sigma^2 = 1$ and

$$k_X(x, y) = k_Y(x, y) = \exp(-\frac{1}{2\sigma^2}||x - y||^2)$$

(Gaussian kernel) as follows.

```
1  k.x=function(x,y)exp(−norm(x−y,"2")^2/2); k.y=k.x
2  n=100
3  for(a in c(0,0.1,0.2,0.4,0.6,0.8)){      ## a: Correlation Coefficients
4      x=rnorm(n); z=rnorm(n); y=a*x+sqrt(1-a^2)*z
5      print(HSIC.1(x,y,k.x,k.y))
6  }
```

We define the HSIC $||m_{XY} - m_X m_Y||^2$ when we test $X \perp\!\!\!\perp Y$ for $m_X = \mathbb{E}_X[k_X(X, \cdot)], m_Y = \mathbb{E}_Y[k_Y(Y, \cdot)]$, and $m_{XY} = \mathbb{E}_{XY}[k_X(X, \cdot)k_Y(Y, \cdot)]$. If we test $X \perp \{Y, Z\}$ between X and $\{Y, Z\}$, then we extend the HSIC to $||m_{XYZ} - m_X m_{YZ}||^2$. For the MMD, we test whether the simultaneous probability of X, Y, Z and the product of the probabilities of X and (Y, Z) are equal. Therefore, we can change $k_Y(y, \cdot)$ to $k_Y(y, \cdot)k_Z(z, \cdot)$. We add arguments to the function HSIC.1 to construct the function HSIC.2 and perform the following operations.

```
1  HSIC.2=function(x,y,z,k.x,k.y,k.z){
2      n=length(x)
3      S=0;for(i in 1:n)for(j in 1:n)S=S+k.x(x[i],x[j])*k.y(y[i],y[j])
       *k.z(z[i],z[j])
4      T=0;
5      for(i in 1:n){
6          T.1=0; for(j in 1:n)T.1=T.1+k.x(x[i],x[j]);
7          T.2=0; for(l in 1:n)T.2=T.2+k.y(y[i],y[l])*k.z(z[i],z[l]);
8          T=T+T.1*T.2
9          }
10     U=0; for(i in 1:n)for(j in 1:n)U=U+k.x(x[i],x[j])
11     V=0; for(i in 1:n)for(j in 1:n)V=V+k.y(y[i],y[j])*k.z(z[i],
12     z[j])
13     return(S/n^2−2*T/n^3+U*V/n^4)
14 }
```

The smaller the value of $\widehat{HSIC}$ is, the more likely independence is, but for random variables X, Y, U, V, the condition $\widehat{HSIC}(X, Y) < \widehat{HSIC}(U, V)$ does not mean that X, Y is closer to independence than U, V. However, in practice, the HSIC is often used as the criterion to measure the certainty of independence.

Example 77 (LiNGAM (Kano and Shimizu [16, 28])) We wish to know the cause-and-effect relation among the random variables X, Y, Z from their independent N realizations x, y, z. For example, we assume that X, Y are generated based on either Model 1 (in which $X = e_1$ and $Y = aX + e_2$ for a constant $a \in \mathbb{R}$ and zero-mean independent variables e_1, e_2) or Model 2 (in which $Y = e_1'$ and $X = a'Y + e_2'$ for a constant $a' \in \mathbb{R}$ and zero-mean independent variables e_1', e_2'). We choose the model with a higher probability between $e_1 \perp\!\!\!\perp e_2$ and $e_1' \perp\!\!\!\perp e_2'$. Then, we can apply the function HSIC.1, where e_2, e_2' are calculated from $y - ax, x - a'y$. For example, using the function

```
1  cc=function(x,y)sum(x*y)/length(x)          ## cc(x,y)/cc(x,x) Partial
2      Correlation Coefficients
3  f=function(u,v)u-cc(u,v)/cc(v,v)*v           ## Residue
```

we can estimate a and a' via f(y,x) and f(x,y), respectively. When we have three variables X, Y, Z, we first determine the upstream variable. To this end, using the function HSIC.2, we compare three independence cases: between x and its residue (f(y,x), f(z,x)), between y and its residue (f(z,y), f(x,y)), and between z and its residue (f(x,z), f(y,z)). For example, if we choose the first pair, then X is the upstream variable.

Then, we choose the midstream variable among the unselected two variables. For example, if X selected in the first round, then we compare two independence sets f(y.x,z.xy) and f(z.x,y.zx). If we use the notation of the program, these are y.x=f(y,x) and z.xy=f(z.x,y.x).

```
1   ## Data Generation ##
2   n=30
3   x=rnorm(n)^2-rnorm(n)^2; y=2*x+rnorm(n)^2-rnorm(n)^2; z=x+y+rnorm(n
        )^2-rnorm(n)^2
4   x=x-mean(x); y=y-mean(y); z=z-mean(z)
5   k.z=k.x
6   ## Estimate UpStream ##
7   cc=function(x,y)sum(x*y)/length(x)
8   f=function(u,v)u-cc(u,v)/cc(v,v)*v
9   x.y=f(x,y); y.z=f(y,z); z.x=f(z,x); x.z=f(x,z); z.y=f(z,y); y.x=
10      f(y,x)
11  v1=HSIC.2(x,y.x,z.x,k.x,k.y,k.z); v2=HSIC.2(y,z.y,x.y,k.y,k.z,k.x)
12      v3=HSIC.2(z,x.z,y.z,k.z,k.x,k.y)
13  if(v1<v2){if(v1<v3)top=1 else top=3} else {if(v2<v3)top=2 else top
        =3} ##
14
15  ## Estimate MidStream ##
16  x.yz=f(x.y,z.y); y.zx=f(y.z,x.z); z.xy=f(z.x,y.x)
17  if(top==1){
18      v1=HSIC.1(y.x,z.xy,k.y,k.z); v2=HSIC.1(z.x,y.zx,k.z,k.y)
```

```
19        if(v1<v2){middle=2; bottom=3} else {middle=3; bottom=2}
20   }
21   if(top==2){
22        v1=HSIC.1(z.y,x.yz,k.z,k.x); v2=HSIC.1(x.y,z.xy,k.x,k.z)
23        if(v1<v2){middle=3; bottom=1} else {middle=1; bottom=3}
24   }
25   if(top==3){
26        v1=HSIC.1(z.y,x.yz,k.z,k.x); v2=HSIC.1(x.y,z.xy,k.x,k.z)
27        if(v1<v2){middle=1; bottom=2} else {middle=2; bottom=1}
28   }
29   ## Output Results ##
30   print(paste("UpStream=",top))
31   print(paste("MidStream=",middle))
32   print(paste("DownStrean=",bottom))
```

In the following, as in the case of the two-sample problem, we construct the distribution under the null hypothesis $X \perp\!\!\!\perp Y$ in two ways.

1. By shifting either $x_1, \ldots, x_N$ or $y_1, \ldots, y_N$ to make X, Y independent, repeatedly obtain the resulting $\widehat{HSIC}$ values to create a histogram that expresses the null hypothesis (permutation test).
2. Compute the (asymptotic) distribution from a statistic whose asymptotic distribution is known (U statistic).

We perform the test by using the procedure shown in the following example.

Example 78 The following procedure randomly rearranges the order of one of the two nonindependent sequences to make them independent, estimates the null distribution of the HSIC values and tests whether they are independent or not (Fig. 5.3).

```
1    ## Enumerate x and show the distribution of HSIC as a histogram ##
2    ## Data Generation ##
3    x=rnorm(n); y=rnorm(n); u=HSIC.1(x,y,k.x,k.y)
4    ## Enumerate x and construct the null hypothesis
5    m=100; w=NULL;
6    for(i in 1:m){x=x[sample(n,n)]; w=c(w,HSIC.1(x,y,k.x,k.y))}
7    ## Set the rejection region
8    v=quantile(w,0.95)
9    ## Graphical Output
10   plot(density(w),xlim=c(min(w,v,u),max(w,v,u)))
11   abline(v=v,col="red",lty=2,lwd=2)
12   abline(v=u,col="blue",lty=1,lwd=2)
```

Now, let us use the unbiased estimate of the HSIC, $\widehat{HSIC}_U$, to find the theoretical asymptotic distribution according to the null hypothesis. Noting that

$$\widehat{HSIC} = \frac{1}{N^4} \sum_{i=1}^{N} \sum_{j=1}^{N} \sum_{q=1}^{N} \sum_{r=1}^{N} h(z_i, z_j, z_q, z_r)$$

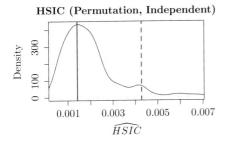
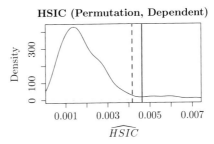

Fig. 5.3 The distribution follows the null hypothesis when using the unbiased estimator $\widehat{HSIC}_U$ of the HSIC. The blue line is the statistic, and the red dotted line is the boundary with the rejection region

for $z_i = (x_i, y_i)$, we have

$$h(z_i, z_j, z_q, z_r) = \frac{1}{4!} \sum_{(t,u,v,w)}^{i,j,h,r} \{k_X(x_t, x_u)k_Y(y_t, y_u) + k_X(x_t, x_u)k_Y(y_v, y_w) - 2k_X(x_t, x_u)k_Y(y_t, y_v)\}$$

where $\sum_{(t,u,v,w)}^{i,j,q,r}$ denotes the sum such that (i, j, q, r) ranges over $(t, u, v, w) = (i, j, h, r)$, i.e., the sum over the permutations of (i, j, h, r). If we modify this estimate to make it an unbiased estimator, we obtain

$$\widehat{HSIC}_U = \frac{1}{\binom{N}{4}} \sum_{i<j<q<r} h(z_i, z_j, z_q, z_r)$$

where $\sum_{i,j,q,r}$ is the sum that ranges over $1 \le i, j, q, r \le N$ without any overlap.

For example, we can construct the program as follows. Since the program consumes memory, the number of samples should be limited to 100 or less. Additionally, since the estimator is different from $\widehat{HSIC}$, it produces different values for the same data. The values of $\widehat{HSIC}_U$ are smaller than those of $\widehat{HSIC}$.

```
h=function(i,j,q,r,x,y,k.x,k.y){
  M=combn(c(i,j,q,r),m=4)
  m=ncol(M)
  S=0
  for(j in 1:m){
    t=M[1,j]; u=M[2,j]; v=M[3,j]; w=M[4,j]
    S=S+k.x(x[t],x[u])*k.y(y[t],y[u])+k.x(x[t],x[u])*k.y(y[v],y[w])
    -2*k.x(x[t],x[u])*k.y(y[t],y[v])
  }
  return(S/m)
}
HSIC.U=function(x,y,k.x,k.y){
M=combn(1:n,m=4)
m=ncol(M)
```

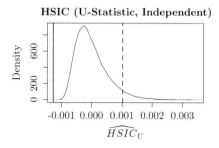
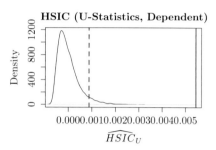

Fig. 5.4 The distribution follows the null hypothesis when using the unbiased estimator of the HSIC, i.e., $\widehat{HSIC}_U$. The blue line is the statistic, and the red dotted line is the boundary with the rejection region. The distribution of the null hypothesis is different from that of the estimator $\widehat{HSIC}$ used in the permutation test. In particular, when X, Y are independent, the unbiased estimator can take a negative value because the true value of the HSIC is zero

```
14   S=0
15   for(j in 1:m)S=S+h(M[1,j],M[2,j],M[3,j],M[4,j],x,y,k.x,k.y)
16   return(S/choose(n,4))
17   }
```

The function $h_1(\cdot)$ is the zero function. For $h_2(\cdot, \cdot)$, we use the following formula.

Proposition 53 (Chwialkowski-Gretton [5]) *Let*

$$\tilde{k}_X(x, x') = k_X(x, x') - \mathbb{E}_{X'}k_X(x, X') - \mathbb{E}_X k_X(X, x') + \mathbb{E}_{XX'}k_X(X, X')$$

and

$$\tilde{k}_Y(y, y') = k_Y(y, y') - \mathbb{E}_{Y'}k_Y(y, Y') - \mathbb{E}_Y k_Y(Y, y') + \mathbb{E}_{YY'}k_Y(Y, Y').$$

Then, $h_2(\cdot, \cdot)$ is given by

$$h_2(z, z') = \frac{1}{6}\tilde{k}_X(x, x')\tilde{k}_Y(y, y')$$

$z = (x, y), z' = (x', y')$.

Proof: The derivation is due to simple transformations. See the original paper for the proof.

Mercer's theorem is not applicable since the kernel h_2 of the integral operator is not nonnegative definite. However, since the kernel is symmetric and its integral operator is self-adjoint, eigenvalues $\{\lambda_i\}$ and eigenfunctions $\{\phi_i\}$ exist (Proposition 27). Therefore, as in the case involving the two-sample problem, the null distribution can be calculated by using Proposition 51. Moreover, the mean of h_2 is zero, i.e., $\tilde{h}_2 = h_2$.

Example 79 Calculate the eigenvalues of the Gram matrix of the positive definite kernel h_2 and divide them by N to obtain the desired eigenvalues (Sect. 3.3). Then, find the distribution that follows the null hypothesis and calculate the rejection region. We construct the following program and execute it. We input a random number that follows a Gaussian distribution with $N = 100$ samples. In Fig. 5.4, the left panel shows a correlation coefficient of 0, and the right panel shows a correlation coefficient of 0.2.

```r
sigma=1; k=function(x,y)exp(-(x-y)^2/sigma^2); k.x=k; k.y=k
## Data Generation
n=100; x=rnorm(n)
a=0          ## Independent
#a=0.2       ## Correlation 0.2
y=a*x+sqrt(1-a**2)*rnorm(n)
# y=rnorm(n)*2   ## The distributions are not equal
## Null Hypothesis
K.x=matrix(0,n,n); for(i in 1:n)for(j in 1:n)K.x[i,j]=k.x(x[i],x[j])
K.y=matrix(0,n,n); for(i in 1:n)for(j in 1:n)K.y[i,j]=k.y(y[i],y[j])
F=array(0,dim=n); for(i in 1:n)F[i]=sum(K.x[i,])/n
G=array(0,dim=n); for(i in 1:n)G[i]=sum(K.y[i,])/n
H=sum(F)/n
I=sum(G)/n
K=matrix(0,n,n)
for(i in 1:n)for(j in 1:n)K[i,j]=(K.x[i,j]-F[i]-F[j]+H)*(K.y[i,j]
-G[i]-G[j]+I)/6
r=20
lambda=eigen(K)$values/n
z=NULL; for(s in 1:10000)z=c(z,1/n*(sum(lambda[1:r]*(rchisq(1:r, df
=1)-1))))
v=quantile(z,0.95)
## Statistics
u=HSIC.U(x,y,k.x,k.y)
## Graphical Output
plot(density(z),xlim=c(min(z,v,u),max(z,v,u)))
abline(v=v,col="red",lty=2,lwd=2)
abline(v=u,col="blue",lty=1,lwd=2)
```

Instead of writing the HSIC as $\|m_{XY} - m_X m_Y\|^2_{H_X \otimes H_Y}$, by using the HS norm, we may write it as $\|\Sigma_{XY}\|^2_{HS}$ or $\|\Sigma_{YX}\|^2_{HS}$. In fact, if $\{e_{X,i}\}$ and $\{e_{Y,j}\}$ are orthonormal bases of H_X, H_Y, respectively, then from the definition of $\|\cdot\|_{HS}$ (Sects. 2.6) and (5.1), we have

$$\|\Sigma_{YX}\|^2_{HS} = \sum_{i=1}^{\infty} \|\Sigma_{YX} e_{X,i}\|^2_{H_Y} = \sum_{i=1}^{\infty}\sum_{j=1}^{\infty} \langle e_{Y,j}, \Sigma_{YX} e_{X,i}\rangle^2_{H_Y}$$

$$= \sum_{i=1}^{\infty}\sum_{j=1}^{\infty} \langle e_{X,i} \otimes e_{Y,j}, m_{XY} - m_X m_Y\rangle^2_{H_X \otimes H_Y} = \|m_{XY} - m_X m_Y\|^2_{H_X \otimes H_Y}.$$

Similarly, $\|\Sigma_{XY}\|^2_{HS}$ has the same value.

5.4   Characteristic and Universal Kernels

Let H and k be an RKHS and its reproducing kernel, respectively. Let $\mathcal{P}$ be the set of distributions that a random variable X follows. Then, we can define the map

$$\mathcal{P} \ni \mu \mapsto \int k(x, \cdot) d\mu(x) \in H$$

which we call the embedding of probabilities in the RKHS.

Since each element of H is generated by $k(x, \cdot)$ $(x \in E)$ or can be written as the limit of its sequence, we describe the condition as

$$\int_E k(x, y) d\mu(x) = 0, \ y \in E \Longrightarrow \mu = 0$$

for $\mu := \mu_1 - \mu_2$. Moreover, $\int_E k(x, y) d\mu(x) = 0$ $(y \in E)$ implies that

$$\int_E \int_E k(x, y) d\mu(x) d\mu(y) = 0. \tag{5.10}$$

If we use $k(x, y) = \phi(x - y) = \int_E e^{i(x-y)w} d\eta(w)$ (Proposition 5), then we can write (5.10) as

$$\int_E | \int_E e^{iwx} d\mu(x)|^2 d\eta(w) = 0.$$

In other words, for the measure η,

$$\hat{\mu}(w) := \int_E e^{iwx} d\mu(x) = 0 \tag{5.11}$$

almost surely holds.

In the following, let η be a finite measure. We call the set of $x \in E$ such that $\eta(U(x, \epsilon)) > 0$ for any $\epsilon > 0$ the support of η and denote it by $E(\eta)$. We note that the support of a finite measure is always a closed set. In fact, for $x \in E \backslash E(\eta)$, if the radius ϵ of the open set $U(x, \epsilon)$ is sufficiently small, then $E(\eta)$, $U(x, \epsilon)$ has no intersection.

Here, if $E(\eta) = E$, then (5.11) means that $\mu = 0$, i.e., $\mu_1 = \mu_2$. On the other hand, if $E(\eta) \subsetneq E$, then a $\mu \neq 0$ exists such that (5.11) holds.

Proposition 54 $k(x, y) = \phi(x - y)$ *is characteristic if and only if the support of the finite measure* η *of* $k(x - y) = \int_E e^{i(x-y)w} d\eta(w)$ *coincides with* E.

For the proof of necessity, see the Appendix at the end of this chapter.

Example 80   The Gaussian kernel in Example 19 and the Laplace kernel in Example 20 are zero-mean Gaussian- and Laplace-distributed, respectively, and the support is the entire interval. Therefore, they are characteristic kernels. On the other hand, the kernel

$$k(x, y) = \phi(x - y)$$

obtained from the characteristic function

$$\phi(t) = \frac{2(1 - \cos(at))}{a^2 t^2}$$

whose probability distribution is a triangular distribution,

$$f(x) = \begin{cases} (1 - \frac{|x|}{a})/a, & |x| < a \\ 0, & \text{Otherwise} \end{cases}$$

is not a characteristic kernel if its support is not equal to E.

Proposition 55 *If the reproducing kernel k_1, k_2 expressed by the bivariate difference between RKHSs H_1, H_2 are both characteristic, then the reproducing kernel $k_1 k_2$ of RKHS $H_1 \otimes H_2$ is also characteristic.*

Proof: If both of $k_1(x_1, y_1) = \phi_1(x_1 - y_1)$, $k_2(x_2, y_2) = \phi_2(x_2 - y_2)$ are characteristic, then the supports of η_1, η_2 are E. Since $k_1(x_1, y_1)k_2(x_2, y_2)$ can be expressed by

$$\phi_1(x_1 - y_1)\phi_2(x_2 - y_2) = \int_E \int_E e^{i(x_1-y_1)^\top w_1} e^{i(x_2-y_2)^\top w_2} d\eta_1(w_1)d\eta_2(w_2)$$

$$= \int_E \int_E e^{i(x_1-y_1, x_2-y_2)^\top (w_1, w_2)} d\eta_1(w_1)d\eta_2(w_2)$$

k_1, k_2 is characteristic as well. □

Let E be a compact set. Suppose that the RKHS H induced by the kernel $k : E \times E \to H$ is a dense (under the uniform norm) subset of the set $C(E)$ of continuous functions $E \to \mathbb{R}$. Then, we say that the kernel k is universal.

To show that the kernel k is universal, we only need to see if the corresponding RKHS satisfies the two Stone-Weierstrass conditions (Proposition 12). Proposition 56 gives a sufficient condition for the kernel k to be universal (see Chap. 2 for the definition of an algebra). However, for practical purposes, Corollary 3 deduced from Proposition 56 is often used.

Proposition 56 (Steinwart [29]) *Let E be compact, and let $k : E \times E \to \mathbb{R}$ be a continuous function with $k(x, x) > 0$ for $x \in E$. If there exists an injective feature map*

$$\Psi : E \ni x \to \Psi(x) = (\Psi_1(x), \Psi_2(x), \ldots) \in l_2 := \{(\alpha_1, \alpha_2, \ldots) \in \mathbb{R}^\infty | \sum_{i=1}^\infty \alpha_j^2 < \infty\}$$

and $A := \text{span}\{\Psi_1, \Psi_2, \ldots\}$ is an algebra, then k is a universal kernel.

Proof: Since $k(x, x) > 0$, $x \in E$, the first condition of Proposition 12 is satisfied. Since $k(\cdot, \cdot)$ is continuous, $\Psi(x) = k(x, \cdot) \in l_2$ is also continuous at each $x \in E$. Moreover, since Ψ is injective, the second condition of Proposition 12 is satisfied, and A is dense in $C(E)$. Furthermore, any $\sum_i \alpha_i \Psi_i(\cdot) \in A$ is also an element of RKHS H with k as the reproducing kernel. Let $\{e_i\}$ be an orthonormal basis of H and $f := \sum_i \alpha_i e_i \in H$. Then, $\langle f(\cdot), \Psi(x) \rangle = f(x)$ holds, which further implies that $\sum_i \alpha_i \Psi_i(x) = \langle \sum_i \alpha_i e_i, \sum_i \Psi_i(x) e_i \rangle = f(x)$. □

Corollary 3 *The infinite-dimensional polynomial kernel (Example 11) is a universal kernel in each compact set of E.*

Proof: The feature map Ψ is injective. Moreover, $A := \text{span}\{\Psi_{m_1, \dots, m_d} | m_1, \dots, m_d \geq 0\}$ is a d-variable polynomial (algebra), and from Proposition 56, the kernel k is universal. □

Example 81 *(Gaussian Kernel)* The exponential-type (Example 6) k_∞ is the universal kernel from Corollary 3. The feature map of the Gaussian kernel (Example 7) is the $\Psi(x)$ of k_∞ divided by $\gamma(x) := k(x, x)^{1/2} > 0$. For $f \in C(E)$, since $\gamma f \in C(E)$, if we let $\|\gamma f(\cdot) - \sum_i \alpha_i \Psi(\cdot)\|_\infty \leq \|\gamma\|_\infty \epsilon$, then we have

$$\|f(\cdot) - \sum_i \alpha_i \gamma^{-1} \Psi(\cdot)\|_\infty \leq \|\gamma\|_\infty^{-1} \|\gamma f(\cdot) - \sum_i \alpha_i \Psi(\cdot)\|_\infty \leq \epsilon$$

Therefore, the Gaussian kernel is universal as well.

The necessary and sufficient condition for Proposition 54 assume that the kernel is a function of the difference between two variables. The following is a sufficient condition, but it refers to kernels in general.

Proposition 57 *A universal kernel on a compact set is characteristic.*

Proof: See the Appendix at the end of the chapter.

Example 82 The Gaussian kernel is characteristic. If a characteristic kernel based on a triangular distribution (Example 80) has a support up to a distance $a > 0$ from the origin and E is a compact set that includes some points outside the support, then its kernel is not universal.

5.5   Introduction to Empirical Processes

In this section, we study a mathematical approach to machine learning called the empirical process. We analyze the accuracy of the MMD estimators by using the Rademacher complexity and concentration inequalities. Through this example, we learn the concept of empirical processes. The derivation performed in this section is based on Gretton et al.'s [11] proof of a proposition regarding the accuracy of the two-sample problem.

In this section, we prove the following proposition. We define the MMD by

$$\sup_{f \in \mathcal{F}} \{\mathbb{E}_P[f(X)] - \mathbb{E}_Q[f(X)]\}$$

where $\mathcal{F}$ is a class of functions. This chapter also deals with the case in which $\mathcal{F} := \{f \in H \,|\, \|f\|_H \leq 1\}$.

Proposition 58 *Suppose that a k_{max} exists such that $0 \leq k(x, y) \leq k_{max}$ for each $x, y \in E$. Then, for any $\epsilon > 0$, we have*

$$\left(|\widehat{MMD}_B^2 - MMD^2| > \frac{4k_{max}}{N} + \epsilon \right) \leq 2 \exp\left(-\epsilon^2 \frac{N}{4k_{max}} \right),$$

where the estimator $\widehat{MMD}_B^2$ of MMD^2 is given by (5.3), and we assume that the number of samples for x, y is equal to N and that $P \neq Q$.

For the proof of Proposition 58, we use an inequality that slightly generalizes Proposition 46.

Proposition 59 (McDiarmid) *Let $f : E^m \to \mathbb{R}$ imply that a $c_i < \infty$ ($i = 1, \cdots, m$) exists satisfying*

$$\sup_{x, x_1, \ldots, x_m} |f(x_1, \ldots, x_m) - f(x_1, \ldots, x_{i-1}, x, x_{i+1}, \ldots, x_m)| \leq c_i.$$

For any probability measure P, $\epsilon > 0$ and $X_1, \ldots, X_m$, we have

$$P\left(f(x_1, \ldots, x_m) - \mathbb{E}_{X_1 \cdots X_m} f(X_1, \ldots, X_m) > \epsilon\right) < \exp\left(-\frac{2\epsilon^2}{\sum_{i=1}^m c_i^2} \right) \quad (5.12)$$

and

$$P\left(|f(x_1, \ldots, x_m) - \mathbb{E}_{X_1 \cdots X_m} f(X_1, \ldots, X_m)| > \epsilon\right) < 2 \exp\left(-\frac{2\epsilon^2}{\sum_{i=1}^m c_i^2} \right). \quad (5.13)$$

Proof: Hereafter, we denote $f(X_1, \cdots, X_N)$ and $\mathbb{E}[f(X_1, \cdots, X_N)]$ by f and $\mathbb{E}[f]$, respectively. If we define

$$V_1 := \mathbb{E}_{X_2 \cdots X_N}[f | X_1] - \mathbb{E}_{X_1 \cdots X_N}[f]$$

$$\vdots$$

$$V_i := \mathbb{E}_{X_{i+1} \cdots X_N}[f | X_1, \cdots, X_i] - \mathbb{E}_{X_i \cdots X_N}[f | X_1, \cdots, X_{i-1}]$$

$$\vdots$$

$$V_N := f - \mathbb{E}_{X_N}[f | X_1, \cdots, X_{N-1}]$$

for $i = 1, i = 2, \cdots, N - 1$, and $i = N$, then we have

$$f - \mathbb{E}_{X_1 \cdots X_N}[f] = \sum_{i=1}^{N} V_i \qquad (5.14)$$

From

$$\mathbb{E}_{X_i}\{\mathbb{E}_{X_{i+1} \cdots X_N}[f|X_1, \cdots, X_i]|X_1, \cdots, X_{i-1}\} = \mathbb{E}_{X_i \cdots, X_N}[f|X_1, \cdots, X_{i-1}]$$

we have

$$\mathbb{E}_{X_i}[V_i|X_1, \cdots, X_{i-1}] = 0. \qquad (5.15)$$

From (5.14), we have

$$f - \mathbb{E}[f] > \epsilon \iff \exp\{t \sum_{i=1}^{N} V_i\} > e^{t\epsilon} \text{ for arbitrary } t > 0.$$

If we apply Markov's inequality (Lemma 6) to the latter equation, then we have

$$P(f - \mathbb{E}[f] \geq \epsilon) \leq \inf_{t>0} e^{-t\epsilon} \mathbb{E}[\exp\{t \sum_{i=1}^{N} V_i\}]. \qquad (5.16)$$

Moreover, from (5.15), we apply Lemma 7 to obtain

$$\mathbb{E}[\exp\{t \sum_{i=1}^{N} V_i\}] = \mathbb{E}_{X_1 \cdots X_{N-1}}[\exp\{t \sum_{i=1}^{N-1} V_i\} \mathbb{E}_{X_N}[\exp\{t V_N\}|X_1, \cdots, X_{N-1}]]$$

$$\leq \mathbb{E}_{X_1 \cdots X_{N-1}}[\exp\{t \sum_{i=1}^{N-1} V_i\}] \exp\{t^2 c_N^2/8\}$$

$$= \exp\{\frac{t^2}{8} \sum_{i=1}^{N} c_i^2\}.$$

Therefore, from (5.16), we have

$$P(f - \mathbb{E}[f] \geq \epsilon) \leq \inf_{t>0} \exp\{-t\epsilon + \frac{t^2}{8} \sum_{i=1}^{N} c_i^2\}.$$

The right-hand side is minimized when $t = 4\epsilon / \sum_{i=1}^{N} c_i^2$, and we obtain (5.12). Replacing f with $-f$, we obtain the other inequality. From both inequalities, we have (5.13).   □

In the following, we denote by

$$\mathcal{F} := \{f \in H \,|\, \|f\|_H \le 1\}$$

the unit ball in the universal (see Sect. 5.4 for the definition of universality) RKHS H w.r.t. a compact E and assume that the kernel of H is less than or equal to k_{max}. Hereafter, let $X_1, \ldots . X_m$ be independent random variables that follow probability P, and let $\sigma_1, \ldots, \sigma_m$ be independent random variables, each of which takes a value of ± 1 equiprobably. Then, we say that the quantity

$$R_N(\mathcal{F}) := \mathbb{E}_\sigma \sup_{f \in \mathcal{F}} |\frac{1}{m} \sum_{i=1}^{m} \sigma_i f(x_i)| \qquad (5.17)$$

is an empirical Rademacher complexity, where $\mathbb{E}_\sigma$ is the operation that takes the expectation w.r.t. $\sigma_1, \ldots, \sigma_m$. If we further take the expectation of (5.17) w.r.t. the probability P, then we call the obtained value $R(\mathcal{F}, P)$ the Rademacher complexity.

Proposition 60 (Bartlett-Mendelson [4]) *Let* $k_{max} := \max_{x, y \in E} k(x, y)$. *Then, we have the following inequality:*

$$R_N(\mathcal{F}) \le \sqrt{\frac{k_{max}}{N}}.$$

In particular, for an arbitrary probability P, *we have*

$$R(\mathcal{F}, P) \le \sqrt{\frac{k_{max}}{N}}.$$

Proof: From $\|f\|_H \le 1$ and $k(x, x) \le k_{max}$, we have

$$R_N(\mathcal{F}) = \mathbb{E}_\sigma [\sup_{f \in \mathcal{F}} |\frac{1}{N} \sum_{i=1}^{N} \sigma_i f(x_i)|] = \mathbb{E}_\sigma [\sup_{f \in \mathcal{F}} |\frac{1}{N} \sum_{i=1}^{N} \sigma_i \langle k(x_i, \cdot), f(\cdot) \rangle_H|]$$

$$= \mathbb{E}_\sigma [\sup_{f \in \mathcal{F}} |\langle f, \frac{1}{N} \sum_{i=1}^{N} \sigma_i k(x_i, \cdot) \rangle_H|]$$

$$\le \mathbb{E}_\sigma [\sup_{f \in \mathcal{F}} \|f\|_H \sqrt{\langle \frac{1}{N} \sum_{i=1}^{N} \sigma_i k(x_i, \cdot), \frac{1}{N} \sum_{i=1}^{N} \sigma_i k(x_i, \cdot) \rangle_H}]$$

$$\le \mathbb{E}_\sigma [\sqrt{\frac{1}{N^2} \sum_{i=1}^{N} \sum_{j=1}^{N} \sigma_i \sigma_j k(x_i, x_j)}] \le \sqrt{\mathbb{E}_\sigma [\frac{1}{N^2} \sum_{i=1}^{N} \sum_{j=1}^{N} \sigma_i \sigma_j k(x_i, x_j)]}$$

$$= \sqrt{\frac{1}{N^2} \sum_{i=1}^{N} \sum_{j=1}^{N} \delta_{i,j} k(x_i, x_j)} \le \sqrt{\frac{k_{max}}{N}}$$

where we use

$$\mathbb{E}[\sigma_i \sigma_j] = \sigma_i^2 \delta_{i,j} = \delta_{i,j}$$

in the derivation. We obtain the other inequality by taking the expectation w.r.t. the probability P.   □

Propositions 59 and 60 are inequalities used for mathematical analysis in machine learning as well as for the proof of Proposition 58.

Proof of Proposition 58: If we define

$$f(x_1, \ldots, x_N, y_1, \ldots, y_N)$$
$$:= \|\frac{1}{N}k(x_1, \cdot) + \ldots + \frac{1}{N}k(x_N, \cdot) - \frac{1}{N}k(y_1, \cdot) - \ldots - \frac{1}{N}k(y_N, \cdot)$$

then from the triangular inequality, we obtain

$$|f(x_1, \ldots, x_N, y_1, \ldots, y_N) - f(x_1, \ldots, x_{j-1}, x, x_{j+1}, \ldots, x_N, y_1, \ldots, y_N)|$$
$$\leq \frac{1}{N}\|k(x_j, \cdot) - k(x, \cdot)\| \leq \frac{2}{N}\sqrt{k_{max}}. \tag{5.18}$$

Next, we obtain the upper bound of the expectation of

$$|MMD^2 - \widehat{MMD}_B^2| = |\sup_{f \in \mathcal{F}}\{\mathbb{E}_P(f) - \mathbb{E}_Q(f)\} - \sup_{f \in \mathcal{F}}\{\frac{1}{N}\sum_{i=1}^N f(x_i) - \frac{1}{N}\sum_{j=1}^N f(y_j)\}|$$
$$\leq \sup_{f \in \mathcal{F}}|\mathbb{E}_P(f) - \mathbb{E}_Q(f) - \{\frac{1}{N}\sum_{i=1}^N f(x_i) - \frac{1}{N}\sum_{j=1}^N f(y_j)\}|.$$

Then, we perform the following derivation:

$$\mathbb{E}_{X,Y} \sup_{f \in \mathcal{F}}|\mathbb{E}_P(f) - \mathbb{E}_Q(f) - \{\frac{1}{N}\sum_{i=1}^N f(X_i) - \frac{1}{N}\sum_{i=1}^N f(Y_i)\}|$$
$$= \mathbb{E}_{X,Y} \sup_{f \in \mathcal{F}}|\mathbb{E}_{X'}\{\frac{1}{N}\sum_{i=1}^N f(X_i') - \frac{1}{N}\sum_{i=1}^N f(X_i)\} - \mathbb{E}_{Y'}\{\frac{1}{N}\sum_{j=1}^N f(Y_j')) - \frac{1}{N}\sum_{j=1}^N f(Y_j)\}|$$
$$\leq \mathbb{E}_{X,Y,X',Y'} \sup_{f \in \mathcal{F}}|\frac{1}{N}\sum_{i=1}^N f(X_i') - \frac{1}{N}\sum_{i=1}^N f(X_i) - \frac{1}{N}\sum_{i=1}^N f(Y_i') + \frac{1}{N}\sum_{i=1}^N f(Y_i)|$$
$$= \mathbb{E}_{X,Y,X',Y',\sigma,\sigma'} \sup_{f \in \mathcal{F}}|\frac{1}{N}\sum_{i=1}^N \sigma_i\{f(X_i') - f(X_i)\} + \frac{1}{N}\sum_{i=1}^N \sigma_i'\{f(Y_i') - f(Y_i)\}|$$
$$\leq \mathbb{E}_{X,X',\sigma} \sup_{f \in \mathcal{F}}|\frac{1}{N}\sum_{i=1}^N \sigma_i\{f(X_i') - f(X_i)\}| + \mathbb{E}_{Y,Y',\sigma'} \sup_{f \in \mathcal{F}}|\frac{1}{N}\sum_{j=1}^n \sigma_j'\{f(Y_j') - f(Y_j)\}|$$
$$\leq 2[R(\mathcal{F}, P) + R(\mathcal{F}, Q)] \leq 2[(k_{max}/N)^{1/2} + (k_{max}/N)^{1/2}] = 4\sqrt{\frac{k_{max}}{N}} \tag{5.19}$$

where the first inequality is due to Jensen's inequality, the second stems from the triangular inequality, the third is derived from the definition of Rademacher complexity, and the fourth due is obtained from the inequality of Rademacher complexity (Proposition 60). From (5.18) and (5.19), for $c_i = \frac{2}{N}\sqrt{k_{max}}$ and $f = MMD^2 - \widehat{MMD}^2$, we have

$$E_{X_1...X_N} f \le 4\sqrt{\frac{k_{max}}{N}}.$$

Finally, we obtain Proposition 58 from Proposition 59.

Hence, Proposition 60 follows from (5.18) and Proposition 59. $\qquad\qquad\square$

Appendix

The essential part of the proof of Proposition 54 was given by Fukumizu [7] but has been rewritten as a concise derivation to make it easier for beginners to understand.

Proof of Proposition 48

The fact that $E \ni x \mapsto k(x, \cdot) \in H$ is measurable means that $\mathbb{E}[k(X, \cdot)]$ can be treated as a random variable. However, the events in $E \times E$ are the direct products of the events generated by each E (the elements of $\mathcal{F} \times \mathcal{F}$). Therefore, if the function $E \times E \ni (x, y) \mapsto k(x, y) \in \mathbb{R}$ is measurable, then the function $E \ni y \mapsto k(x, y) \in \mathbb{R}$ is measurable for each $x \in E$ (even if $y \in E$ is fixed, $(x, y) \mapsto k(x, y)$ is still measurable). In the following, we show that any function belonging to H is measurable. First, we note that $H_0 = \text{span}\{k(x, \cdot)|x \in E\}$ is dense in H. Additionally, we note that for the sequence $\{f_n\}$ in H_0, $\|f - f_n\|_H \to 0$ ($n \to \infty$) means that $|f(x) - f_n(x)| \to 0$ for each $x \in E$ (Proposition 35). The following lemma implies that f is measurable.

Lemma 8 *If $f_n : E \to \mathbb{R}$ is measurable and $f_n(x)$ converges to $f(x)$ for each $x \in E$, then $f : E \to \mathbb{R}$ is also measurable.*

Proof: The proof follows after the proof of this proposition.

We assume that Lemma 8 is valid. We define the measurability of $\Psi : E \ni x \mapsto k(x, \cdot) \in H$ by

$$\{x \in E \mid \|f - k(x, \cdot)\|_H < \delta\} \in \mathcal{F}$$

for any $f \in H$ and $\delta > 0$ (this is an extension to the case where $H = \mathbb{R}$). Moreover, we have

$$\|f - k(x, \cdot)\|_H < \delta \iff k(x, x) - 2f(x) < \delta^2 - \|f\|_H^2.$$

In addition, since $k(\cdot, \cdot)$ is measurable, $E \ni x \mapsto k(x, x) \in \mathbb{R}$ is also measurable. Moreover, since $f(x)$ is measurable, so is $k(x, x) - 2f(x)$. Thus, Ψ is measurable.

$\square$

Proof of Lemma 8

It is sufficient to show that $f^{-1}(B) \in \mathcal{F}$ for any open set B. We fix $B \subseteq \mathbb{R}$ arbitrarily and let $F_m := \{y \in B | U(y, 1/m) \subseteq B\}$, where $U(y, r) := \{x \in \mathbb{R} \mid d(x, y) < r\}$. From the definition, we have the following two equations.

$$f(x) \in B \iff \text{for some } m, \ f(x) \in F_m$$

$$f(x) \in F_m \iff \text{for some } k, \ f_n(x) \in F_m, \ n \geq k.$$

In other words, we have

$$f^{-1}(B) = \cup_m f^{-1}(F_m) = \cup_m \cup_k \cap_{n \geq k} f_n^{-1}(F_m) \in \mathcal{F}.$$

$\square$

Proof of Proposition 49

The evaluation is finite for arbitrary $g = \sum_{i=1}^{\infty} \sum_{j=1}^{\infty} e_{X,i} e_{Y,j} \in H_X \otimes H_Y$ and $(x, y) \in E$. In fact, we have

$$|g(x, y)| \leq \sum_{i=1}^{\infty} \sum_{j=1}^{\infty} |a_{i,j}| \cdot |e_{X,i}(x)| \cdot |e_{Y,j}(y)| \leq \sum_{i=1}^{\infty} |e_{X,i}(x)| \cdot \left(\sum_{j=1}^{\infty} e_{Y,j}^2(y) \right)^{1/2} \left(\sum_{j=1}^{\infty} a_{i,j}^2(y) \right)^{1/2} \tag{5.20}$$

where we apply Cauchy-Schwarz's inequality (2.5) to $\sum_{j=1}^{\infty}$. If we set $k_Y(y, \cdot) = \sum_j h_j(y) e_{Y,j}(\cdot)$, then from $\langle e_{Y,i}(\cdot), k_Y(y, \cdot) \rangle = e_{Y,i}(y)$, we have $h_i(y) = e_{Y,i}(y)$ and $k_Y(y, \cdot) = \sum_{j=1}^{\infty} e_{Y,j}(y) e_{Y,j}(\cdot)$. Thus, we obtain

$$\sum_{j=1}^{\infty} e_{Y,j}^2(y) = k_Y(y, y) \tag{5.21}$$

and

$$\sum_{i=1}^{\infty} |e_{X,i}(x)| \cdot \left(\sum_{j=1}^{\infty} a_{i,j}^2 \right)^{1/2} \leq \left(\sum_{i=1}^{\infty} e_{X,i}^2(x) \right)^{1/2} \left(\sum_{i=1}^{\infty} \sum_{j=1}^{\infty} a_{i,j}^2 \right)^{1/2} = \sqrt{k_X(x, x)} \|g\| \tag{5.22}$$

where we apply Cauchy-Schwarz's inequality (2.5) to $\sum_{i=1}^{\infty}$. Note that (5.20), (5.21), and (5.22) imply that $|g(x, y)| \leq \sqrt{k_X(x, x)} \sqrt{k_Y(y, y)} \|g\|$. Thus, $H_X \otimes H_Y$ is an RKHS.

From $k_X(x, \cdot) \in H_X$, $k_Y(y, \cdot) \in H_Y$, we have that $k(x, \cdot, y, \star) := k_X(x, \cdot)$
$k_Y(y, \cdot) \in H_X \otimes H_Y$ for $k(x, x', y, y') := k_X(x, x')k_Y(y, y')$. From

$$
\begin{aligned}
g(x, y) &= \sum_{i=1}^{\infty} \sum_{j=1}^{\infty} a_{i,j} e_{X,i}(x) e_{Y,j}(y) \\
&= \sum_{i=1}^{\infty} \sum_{j=1}^{\infty} a_{i,j} \langle e_{X,i}(\cdot), k_X(x, \cdot) \rangle_{H_X} \langle e_{Y,i}(\star), k_Y(y, \star) \rangle_{H_Y} \\
&= \sum_{i=1}^{\infty} \sum_{j=1}^{\infty} a_{i,j} \langle e_{X,i}(\cdot) e_{Y,j}(\star), k(x, \cdot, y, \star) \rangle_H \\
&= \langle \sum_{i=1}^{\infty} \sum_{j=1}^{\infty} a_{i,j} e_{X,i}(\cdot) e_{Y,j}(\cdot), k(x, \cdot, y, \star) \rangle_H \\
&= \langle g(\cdot, \star), k(x, \cdot, y, \star) \rangle
\end{aligned}
$$

k is the reproducing kernel of $H_X \otimes H_Y$. $\square$

Proof of Proposition 54 (Necessity)

Let W be an open set centered at the origin with a radii of $\epsilon > 0$ and $w_0 \in E$. We assume that $w_0 + W$ has a measure of 0 and show that this contradicts another assumption, i.e., that $k(x, y) = \phi(x - y)$ is a characteristic kernel. In this case, η is an even function, and $-w_0 + W$ is also of measure 0 ($\pm w_0 + W \subseteq E \backslash E(\eta)$), where we use the fact that $g(w) := (\epsilon - \|w\|_2)_+^{(d+1)/2}$ is nonnegative definite when $E = \mathbb{R}^d$ ($d \geq 1$) (see [8] for the proof). From Proposition 5 (Bochner's Theorem), there exists a finite measure μ such that $g(w) = \int_E e^{iw^\top x} \mu(x)$. Moreover, the closure of $\pm w_0 + W$ is the support of

$$
h(w) = g(w - w_0) + g(w + w_0) = \int_E e^{iw^\top x} 2\cos(w_0^\top x) d\mu(x).
$$

Since the support of h has no intersection with $E(\eta)$ and $\pm w_0 \notin W$, we have $h(0) = 0$. Therefore, we obtain $\nu(E) = 0$ for

$$
\nu(B) := \int_B 2\cos(w_0 x) d\mu(x), \quad B \in \mathcal{F}.
$$

Since g is not zero, ν is not the zero measure. Thus, using the total variation

$$
|\nu|(B) := \sup_{\cup B_i = B} \sum_{i=1}^{n} |\nu(B_i)|, \quad B \in \mathcal{F}
$$

where sup is the supremum when dividing $\mathcal{F}$ into $B_i \in \mathcal{F}$, we define the constant $c := |\nu|(E)$ and the finite measures $\mu_1 := \frac{1}{c}|\nu|$ and $\mu_2 := \frac{1}{c}\{|\nu| - \nu\}$. From $\nu(E) = 0$, we observe that μ_1 and μ_2 are both probabilities and that $\mu_1 \neq \mu_2$. Additionally,

Fig. 5.5 Proof of Proposition 57. As n grows, the slope of f_n rapidly increases at the border of F, U. Therefore, if $\int_E f_n dP = \int_E f_n dQ$ for all $\{f_n\}$, we require $P = Q$ (Dudley "Real Analysis and Probability"[6])

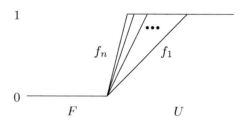

we have

$$c(d\mu_1 - d\mu_2) = dv = 2\cos(w_0 x)d\mu.$$

From Fubini's theorem, we can write the difference between the expectations w.r.t. probabilities μ_1, μ_2 as

$$\int_E \phi(x-y)d\mu_1(y) - \int_E \phi(x-y)d\mu_2(y) = \frac{1}{c}\int_E \phi(x-y)2\cos(w_0^\top y)d\mu(y)$$

$$= \frac{1}{c}\int 2\cos(w_0^\top y)\int e^{i(x-y)^\top w}d\eta d\mu(y) = \frac{1}{c}\int_E e^{ix^\top w}h(w)d\eta(w).$$

However, since the supports of h and η do not intersect, the value is zero, which contradicts the assumption that $\phi(x-y)$ is a characteristic kernel. □

Proof of Proposition 57

For any bounded continuous f, if $\int_E f dP = \int_E f dQ$ holds, this implies that $P = Q$ (Fig. 5.5). In fact, let U be an open subset of E, and let V be its complement. Furthermore, let $d(x, V) := \inf_{y \in V} d(x, y)$ and $f_n(x) := \min(1, nd(x, V))$. Then, f_n is a bounded continuous function on E, and $f_n(x) \leq I(x \in U)$ and $f_n(x) \to I(x \in U)$ as $n \to \infty$ for each $x \in \mathbb{R}$; Thus, by the monotonic convergence theorem, $\int_E f_n dP \to P(U)$ and $\int_E f_n dQ \to Q(U)$ hold. By our assumption, $\int_E f_n dP = \int_E f_n dQ$ and $P(U) = Q(U)$, i.e., $P(V) = Q(V)$ holds[2] In other words, every event is guaranteed to be a closure event.. Let E be a compact set. For each element $g \in H$ in the RKHS H of the universal kernel, the same argument follows since $\sup_{x \in E} |f(x) - g(x)|$ can be arbitrarily small for any $f \in C(E)$. That is, if $\int g dP = \int g dQ$ holds for any $g \in H$, then $P = Q$, so the universal kernel is characteristic. □

Exercises 65~83

65. Proposition 49 can be derived according to the following steps. Which part of the proof in the Appendix does each step correspond to?

---

[2] If E is compact, then for any $A \in \mathcal{F}$, $P(A) = \{P(V)|V$ is a closed set, $V \subseteq A, V \in \mathcal{V}\}$ (Theorem 7.1.3, Dudley 2003).

(a) Show that $|g(x, x, y, y)| \leq \sqrt{k_X(x, x)}\sqrt{k_Y(y, y)}\|g\|$ for $g \in H_X \otimes H_Y$ and $x \in E_X, y \in E_Y$ (from Proposition 33, this implies that H is some RKHS).

(b) Show that $k(x, \cdot, y, \star) := k_X(x, \cdot)k_Y(y, \star) \in H$ when $x \in E_X, y \in E_Y$ are fixed.

(c) Show that $f(x, y) = \langle f(\cdot, \star), k(x, \cdot, y, \star)\rangle_H$.

66. How can we define the average of the elements of $H_{X \oplus Y}$, $m_{XY} = \mathbb{E}_{XY}[k_X(\cdot) k_Y(\cdot)]$? Define the average in the same way that we defined m_X using Riesz's lemma (Proposition 22).

67. Show that $\Sigma_{YX} \in B(H_X, H_Y)$ exists such that

$$\langle fg, m_{XY} - m_X m_Y\rangle_{H_X \otimes H_Y} = \langle \Sigma_{YX} f, g\rangle_{H_Y}$$

for each $f \in H_X, g \in H_Y$.

68. The MMD is generally defined as $\sup_{f \in \mathcal{F}}\{\mathbb{E}_P[f(X)] - \mathbb{E}_Q[f(X)]\}$ for some set $\mathcal{F}$ of functions. Assuming that $\mathcal{F} := \{f \in H \mid \|f\|_H \leq 1\}$, show that the MMD is $\|m_P - m_Q\|_H$. Furthermore, show that we can transform the MMD as follows.

$$\text{MMD}^2 = \mathbb{E}_{XX'}[k(X, X')] + \mathbb{E}_{YY'}[k(Y, Y')] - 2\mathbb{E}_{XY}[k(X, Y)]$$

where X' and X (Y' and Y) are independent random variables that follow the same distribution.

69. Show that the squared MMD estimator (5.4) is unbiased.

70. In the two-sample problem solved by a permutation test in Example 71, for the case when the numbers of samples are m, n (can be different) instead of the same n and m, n are both even numbers, modify the entire program in Example 71 to examine whether it works correctly ($m = n$ in Example 71).

71. For the function h in (5.6), show that h_1 is a function that always takes a value of zero and that $\tilde{h}_2$ and h coincide as functions.

72. Show that the fact that random variables X, Y that follow Gaussian distributions are independent is equivalent to the condition that their correlation coefficient is zero. Additionally, give an example of two variables whose correlation coefficient is zero but that are not independent.

73. Prove the following equation.

$$\|m_{XY} - m_X m_Y\|^2 = \mathbb{E}_{XX'YY'}[k_X(X, X')k_Y(Y, Y')]$$
$$-2\mathbb{E}_{XY}\{\mathbb{E}_{X'}[k_X(X, X')]\mathbb{E}_{Y'}[k_Y(Y, Y')]\} + \mathbb{E}_{XX'}[k_X(X, X')]\mathbb{E}_{YY'}[k_Y(Y, Y')]$$

74. Show that the HSIC estimator

$$\widehat{HSIC} := \frac{1}{N^2}\sum_i\sum_j k_X(x_i, x_j)k_Y(y_i, y_j) - \frac{2}{N^3}\sum_i\sum_j k_X(x_i, x_j)\sum_h k_Y(y_i, y_h)$$
$$+ \frac{1}{N^4}\sum_i\sum_j k_X(x_i, x_j)\sum_h\sum_r k_Y(y_h, y_r)$$

can be written as $\widehat{HSIC} = \text{trace}(K_X H K_Y H)$ using $K_X = (k_X(x_i, x_j))_{i,j}$, $K_Y = (k_Y(y_i, y_j))_{i,j}$, and $H = I - \frac{1}{N}E$, where $I \in \mathbb{R}^{N \times N}$ is the unit matrix and $E \in \mathbb{R}^{N \times N}$ is the matrix such that all the elements are ones. Additionally, construct R programs for each computation. Moreover, examine that both output the same results for the Gaussian kernels k.x and k.y with $\sigma^2 = 1$; generate random numbers for the standard Gaussian variables X and Y whose correlations are $a = 0, 0.1, 0.2, 0.4, 0.6, 0.8$.

75. When we test the independence $X \perp\!\!\!\perp \{Y, Z\}$ of X and $\{Y, Z\}$, the HSIC is extended as $\|m_{XYZ} - m_X m_{YZ}\|^2$. That is, we can transform $k_Y(y, \cdot)$ into $k_Y(y, \cdot)k_Z(z, \cdot)$. Construct the function HSIC.2 by adding arguments to the function HSIC.1; generate a random number according to $X \perp\!\!\!\perp \{Y, Z\}$, and verify that the obtained value is sufficiently small.

76. Utilizing the class of LiNGAM and the function

```
cc=function(x,y)sum(x*y)/length(x)
f=function(u,v)u-cc(u,v)/cc(v,v)*v
```

we wish to estimate whether each variable X, Y, Z is either upstream, midstream, or downstream. Fill in the blanks by generating random numbers X, Y, Z that do not follow the Gaussian distribution, and estimate which variables among X, Y, Z are upstream, midstream, and downstream from the random numbers alone.

```
## Estimate UpStream ##
cc=function(x,y)sum(x*y)/length(x)
f=function(u,v)u-cc(u,v)/cc(v,v)*v
x.y=f(x,y); y.z=f(y,z); z.x=f(z,x); x.z=f(x,z); z.y=f(z,y); y.x=
    f(y,x)
v1=HSIC.2(x,y.x,z.x,k.x,k.y,k.z); v2=HSIC.2(y,z.y,x.y,k.y,k.z,k.
    x)
    v3=HSIC.2(z,x.z,y.z,k.z,k.x,k.y)
if(v1<v2){if(v1<v3)top=1 else top=3} else {if(v2<v3)top=2 else
    top=3} ##

## Estimate MidStream ##
x.yz=f(x.y,z.y); y.zx=f(y.z,x.z); z.xy=f(z.x,y.x)
if(top==1){
    v1= ## Blank(1) ##; v2= ## Blank(2) ##
    if(v1<v2){middle=2; bottom=3} else {middle=3; bottom=2}
}
if(top==2){
    v1= ## Blank(3) ##; v2= ## Blank(4) ##
    if(v1<v2){middle=3; bottom=1} else {middle=1; bottom=3}
}
if(top==3){
    v1= ## Blank(5) ##; v2= ## Blank(6) ##
    if(v1<v2){middle=1; bottom=2} else {middle=2; bottom=1}
}
## Output Results ##
print(paste("UpStream=",top))
```

```
25  print(paste("MidStream=",middle))
26  print(paste("DownStream=",bottom))
```

77. We wish to make two sequences independent by shifting one of $x_1, \ldots, x_N$ or $y_1, \ldots, y_N$, and then we want to repeat the process of calculating $\widehat{HSIC}$. We wish to create a histogram that expresses a distribution that follows the null hypothesis. For this purpose, we constructed the following program. Why can we obtain the null hypothesis $(X, Y$ are independent) by permutation? Where in the program do we obtain the HSIC statistics, and where do we obtain the multiple HSIC values that follow the null hypothesis?

```
1  x=rnorm(n); y=rnorm(n); u=HSIC.1(x,y,k.x,k.y)
2  m=100; w=NULL;
3  for(i in 1:m){x=x[sample(n,n)]; w=c(w,HSIC.1(x,y,k.x,k.y))}
4  v=quantile(w,0.95)
5  plot(density(w),xlim=c(min(w,v,u),max(w,v,u)))
6  abline(v=v,col="red",lty=2,lwd=2)
7  abline(v=u,col="blue",lty=1,lwd=2)
```

78. In the MMD (Sect. 5.2) and HSIC (Sect. 5.3), we cannot apply Mercer's theorem because the kernel of the integral operator is not nonnegative definite. However, in both cases, the integral operator possesses eigenvalues and eigenfunctions. Why?

79. Show that $k(x, y) = \phi(x - y)$, $\phi(t) = e^{-|t|}$ is a characteristic kernel.

80. In the proof of Proposition 54 (necessity, Appendix), we used the fact that $g(w) := (\epsilon - \|w\|_2)_+^{(d+1)/2}$ is nonnegative definite [8]. Verify that this fact is correct for $d = 1$ by proving the following equality.

$$\frac{1}{2\pi} \int_{-\epsilon}^{\epsilon} g(w)e^{-iwx}\,dw = \frac{1 - \cos(x\epsilon)}{\pi x^2}.$$

81. Why is the exponential type a universal kernel? Why is the characteristic kernel based on a triangular distribution not a universal kernel?

82. Explain why the three equations and four inequalities hold in the following derivation of the upper bound on the Rademacher complexity.

$$
R_N(\mathcal{F}) = \mathbb{E}_\sigma [\sup_{f \in \mathcal{F}} |\frac{1}{N} \sum_{i=1}^{N} \sigma_i f(x_i)|] = \mathbb{E}_\sigma [\sup_{f \in \mathcal{F}} |\frac{1}{N} \sum_{i=1}^{N} \sigma_i \langle k(x_i, \cdot), f(\cdot) \rangle_H |]
$$

$$
= \mathbb{E}_\sigma [\sup_{f \in \mathcal{F}} |\langle f, \frac{1}{N} \sum_{i=1}^{N} \sigma_i k(x_i, \cdot) \rangle_H |]
$$

$$
\leq \mathbb{E}_\sigma [\sup_{f \in \mathcal{F}} \|f\|_H \sqrt{\langle \frac{1}{N} \sum_{i=1}^{N} \sigma_i k(x_i, \cdot), \frac{1}{N} \sum_{i=1}^{N} \sigma_i k(x_i, \cdot) \rangle_H}]
$$

$$
\leq \mathbb{E}_\sigma [\sqrt{\frac{1}{N^2} \sum_{i=1}^{N} \sum_{j=1}^{N} \sigma_i \sigma_j k(x_i, x_j)}]
$$

$$
\leq \sqrt{\mathbb{E}_\sigma [\frac{1}{N^2} \sum_{i=1}^{N} \sum_{j=1}^{N} k(x_i, x_j)]} \leq \sqrt{\frac{k_{max}}{N}}
$$

83. Explain why the one equality and four inequalities hold for the derivation of the upper bound of $|MMD^2 - \widehat{MMD}_B^2|$ below.

$$
\mathbb{E}_{X,Y} \sup_{f \in \mathcal{F}} |\mathbb{E}_{X'}\{\frac{1}{N} \sum_{i=1}^{N} f(x_i') - \frac{1}{N} \sum_{i=1}^{N} f(x_i)\} - \mathbb{E}_{Y'}\{\frac{1}{N} \sum_{j=1}^{N} f(y_j')) - \frac{1}{N} \sum_{j=1}^{N} f(y_j)\}|
$$

$$
\leq \mathbb{E}_{X,Y,X',Y'} \sup_{f \in \mathcal{F}} |\frac{1}{N} \sum_{i=1}^{N} f(x_i') - \frac{1}{N} \sum_{i=1}^{N} f(x_i) - \frac{1}{N} \sum_{i=1}^{N} f(y_i') + \frac{1}{N} \sum_{i=1}^{N} f(y_i)|
$$

$$
= \mathbb{E}_{X,Y,X',Y',\sigma,\sigma'} \sup_{f \in \mathcal{F}} |\frac{1}{N} \sum_{i=1}^{N} \sigma_i \{f(x_i') - f(x_i)\} + \frac{1}{N} \sum_{i=1}^{N} \sigma_i' \{f(y_i') - f(y_i)\}|
$$

$$
\leq \mathbb{E}_{X,X',\sigma} \sup_{f \in \mathcal{F}} |\frac{1}{N} \sum_{i=1}^{N} \sigma_i \{f(x_i') - f(x_i)\}| + \mathbb{E}_{Y,Y',\sigma'} \sup_{f \in \mathcal{F}} |\frac{1}{N} \sum_{j=1}^{n} \sigma_j' \{f(y_j') - f(y_j)\}|
$$

$$
\leq 2[R(\mathcal{F}, P) + R(\mathcal{F}, Q)]
$$

$$
\leq 2[(k_{max}/N)^{1/2} + (k_{max}/N)^{1/2}]
$$

Chapter 6
Gaussian Processes and Functional Data Analyses

A stochastic process may be defined either as a sequence of random variables $\{X_t\}_{t \in T}$, where T is a set of times, or as a function $X_t(\omega) : T \to \mathbb{R}$ of $\omega \in \Omega$. We define a Gaussian process as a stochastic process $\{X_t\}$ such that X_t $(t \in T')$ follows a multivariate Gaussian distribution for any finite subset T' of T. In this chapter, we generalize the one-dimensional T to a multidimensional set E for the consideration of Gaussian processes. We mainly deal with the variations of $\omega \in \Omega$ in $f(\omega, x)$, while thus far, we have dealt with the variations of $x \in E$ in $f(\omega, x)$. The Gaussian process has been applied to various aspects of machine learning. We examine the relation between Gaussian processes and kernels. The chapter's first half consists of regression, classification, and computational reduction treatments, and the last part studies the Karhunen-Lóeuvre expansion and its surrounding theory. Finally, we study functional data analyses, which are closely related to stochastic processes.

6.1 Regression

Let E and $(\Omega, \mathcal{F}, \mu)$ be a set and a probability space. If the correspondence between $\Omega \ni \omega \mapsto f(\omega, x) \in \mathbb{R}$ is measurable for each $x \in E$, i.e., if $f(\omega, x)$ is a random variable at each $x \in E$, then we say that $f : \Omega \times E \to \mathbb{R}$ is a stochastic process. Moreover, if the random variables $f(\omega, x_1), \ldots, f(\omega, x_N)$ follow an N-variable Gaussian distribution for any $N \geq 1$ and any finite number of elements $x_1, \ldots, x_N \in E$, then we call f a Gaussian process. We define the covariance between $x_i, x_j \in E$ by

$$\int_\Omega \{f(\omega, x_i) - m(x_i)\}\{f(\omega, x_j) - m(x_j)\}d\mu(\omega)$$

where $m(x) := \int_\Omega f(\omega, x)d\mu(\omega)$ is the expectation of $f(\omega, x)$ for $x \in E$. Then, no matter what N and $x_1, \ldots, x_N$ we choose, their covariance matrices are nonnegative

© The Author(s), under exclusive license to Springer Nature Singapore Pte Ltd. 2022
J. Suzuki, *Kernel Methods for Machine Learning with Math and R*,
https://doi.org/10.1007/978-981-19-0398-4_6

definite. Thus, we can write the covariance matrix by using a positive definite kernel $k : E \times E \to \mathbb{R}$. Therefore, the Gaussian process can be uniquely expressed in terms of a pair (m, k) containing the mean $m(x)$ of each $x \in E$ and the covariance $k(x, x')$ of each $(x, x') \in E \times E$.

In general, a random variable is a map of $\Omega \to \mathbb{R}$, and we should make ω explicit, i.e., $f(\omega, x)$, but for simplicity, for the time being, we make ω implicit, i.e., $f(x)$, even if it is a random variable.

Example 83 Let $m_X \in \mathbb{R}^N$ and $k_{XX} \in \mathbb{R}^{N \times N}$ be the mean and covariance matrix, respectively, of the Gaussian process (m, k) at $x_1, \ldots, x_N \in E := \mathbb{R}$. In general, for a mean μ and a covariance matrix $\Sigma \in \mathbb{R}^{N \times N}$, Σ is nonnegative definite, and there exists a lower triangular matrix $R \in \mathbb{R}^{N \times N}$ with $\Sigma = R R^\top$ (Cholesky decomposition). Therefore, to generate random numbers that follow $N(m_X, k_{XX})$ from N independent random numbers $u_1, \ldots, u_N$ that follow the standard Gaussian distribution, we can calculate $f_X := R_X u + m_X \in \mathbb{R}^N$ for $k_{XX} := R_X R_X^\top$ with $u = [u_1, \ldots, u_N]$. In fact, the expectation and the covariance matrix of f_X are m_X and

$$\mathbb{E}[(f_X - m_X)(f_X - m_X)^\top] = \mathbb{E}[R_X u u^\top R_X^\top] = R_X \mathbb{E}[u u^\top] R_X^\top = R_X R_X^\top = k_{XX},$$

respectively. This procedure can be described in the R language as follows.

```
## Definitions of (m,k)
m=function(x) 0; k=function(x,y) exp(-(x-y)^2/2)
## Definition of Function gp.sample
gp.sample=function(x,m,k){
  n=length(x)
  m.x=m(x)
  k.xx=matrix(0,n,n); for(i in 1:n)for(j in 1:n)k.xx[i,j]=k(x[i],
  x[j])
  R=t(chol(k.xx))
  u=rnorm(n)
  return(as.vector(R%*%u+m.x))
}
## Generate random numbers and its covariance matrix to compare with k.xx
x=seq(-2,2,1); n=length(x)
r=100; z=matrix(0,r,n); for(i in 1:r)z[i,]=gp.sample(x,m,k)
k.xx=matrix(0,n,n); for(i in 1:n)for(j in 1:n)k.xx[i,j]=k(x[i],x[j])
```

```
> cov(z)
        [,1]         [,2]         [,3]          [,4]            [,5]
[1,]   1.1340978 0.680820479 0.1016659 -0.151665584 -0.113949224
[2,]   0.6808205 0.955260633 0.4835023  0.007489523  0.009933267
[3,]   0.1016659 0.483502265 1.0569062  0.692036816  0.258458530
[4,]  -0.1516656 0.007489523 0.6920368  1.122445813  0.703487844
[5,]  -0.1139492 0.009933267 0.2584585  0.703487844  0.959871890
> k.xx
```

```
          [,1]        [,2]       [,3]      [,4]          [,5]
[1,]  1.0000000000 0.6065307 0.1353353 0.0111090 0.0003354626
[2,]  0.6065306597 1.0000000 0.6065307 0.1353353 0.0111089965
[3,]  0.1353352832 0.6065307 1.0000000 0.6065307 0.1353352832
[4,]  0.0111089965 0.1353353 0.6065307 1.0000000 0.6065306597
[5,]  0.0003354626 0.0111090 0.1353353 0.6065307 1.0000000000
```

In the R language, the output is R such that the Cholesky decomposition is $R^\top R$ rather than $R R^\top$.

In general, E does not have to be $\mathbb{R}$. Gaussian processes are a class of stochastic processes, and we might have the impression that the set E is the entire real number set or a subset of it, but in fact, there is no further restriction as long as we define the positive definite kernel k on $E \times E$. Once we choose (m, k), we generate N-variate Gaussian random variables according to (m, k), regardless of the selected E.

Example 84 For $E = \mathbb{R}^2$, we can similarly obtain random numbers that follow the N-variate multivariate Gaussian distribution.

```
1   ## Definitions of (m,k)
2   m=function(x) x[,1]-x[,2]
3   k=function(x,y) exp(-sum((x-y)^2)/2)
4   ## Definition of function gp.sample
5   gp.sample=function(x,m,k){
6     n=nrow(x)
7     m.x=m(x)
8     k.xx=matrix(0,n,n); for(i in 1:n)for(j in 1:n)k.xx[i,j]=k(x[i,],x[
        j,])
9     R=t(chol(k.xx))
10    u=rnorm(n)
11    return(R%*%u+m.x)
12  }
13  ## Generate random numbers and covariance matrix to compare with k.xx
14  n=5; x=matrix(rnorm(n*2),n,2)
15  r=100; z=matrix(0,r,n); for(i in 1:r)z[i,]=gp.sample(x,m,k)
16  k.xx=matrix(0,n,n); for(i in 1:n)for(j in 1:n)k.xx[i,j]=k(x[i,],x[j
        ,])
```

In the above procedure, we have $N = n$ and $p = n$. Regardless of the dimension p of E, once we determine m.x and k.xx, then gp.sample generates N-variable Gauss samples.

```
> cov(z)
          [,1]       [,2]       [,3]       [,4]       [,5]
[1,]  0.9442610 0.6247434 0.34650892 0.30256420 0.21339075
[2,]  0.6247434 0.8704980 0.21572575 0.17389875 0.63640507
[3,]  0.3465089 0.2157258 1.11319581 1.10004943 0.09225664
[4,]  0.3025642 0.1738987 1.10004943 1.09753987 0.05793784
[5,]  0.2133908 0.6364051 0.09225664 0.05793784 0.87728064
> k.xx
```

	[,1]	[,2]	[,3]	[,4]	[,5]
[1,]	1.0000000	0.7415920	0.9995366	0.9999418	0.3170654
[2,]	0.7415920	1.0000000	0.7589062	0.7477614	0.7591127
[3,]	0.9995366	0.7589062	1.0000000	0.9998068	0.3318859
[4,]	0.9999418	0.7477614	0.9998068	1.0000000	0.3222743
[5,]	0.3170654	0.7591127	0.3318859	0.3222743	1.0000000

Then, as in the usual regression procedure, we assume that $x_1, \ldots, x_N \in E$ and $y_1, \ldots, y_N \in \mathbb{R}$ are generated according to

$$y_i = f(x_i) + \epsilon_i \tag{6.1}$$

through the use of an unknown function $f : E \to \mathbb{R}$, where ϵ_i follows a Gaussian distribution with a mean of 0 and a variance of σ^2 and is independent for each $i = 1, \ldots, N$. The likelihood is

$$\prod_{i=1}^{N} [\frac{1}{\sqrt{2\pi\sigma^2}} \exp\{-\frac{(y_i - f(x_i))^2}{2\sigma^2}\}]$$

when the function f is known (fixed). In the following, we assume that the function f randomly varies, and we regard the Gaussian process (m, k) as its prior distribution. That is, we consider the model $f_X \sim N(m_X, k_{XX})$ with $y_i | f(x_i) \sim N(f(x_i), \sigma^2)$ as $f_X = (f(x_1), \ldots, f(x_N))$. Then, we calculate the posterior distribution of $f(z_1), \ldots, f(z_n)$ corresponding to $z_1, \ldots, z_n \in E$, which is different from $x_1, \ldots, x_N$. The variations in $y_1, \ldots, y_N$ is due to the variations in f and ϵ_i. Thus, the covariance matrix is

$$k_{XX} + \sigma^2 I = (k(x_i, x_j) + \sigma^2 \delta_{i,j})_{i,j=1,\ldots,N} \in \mathbb{R}^{N \times N}.$$

On the other hand, the variation in $f(z_1), \ldots, f(z_n)$ is due only to the variation in f. Therefore, the covariance matrix is $k_{ZZ} = (k(z_i, z_j))_{i,j=1,\ldots,n} \in \mathbb{R}^{n \times n}$. Moreover, the variances of y_i and $f(z_j)$ are those of $f(x_i)$ and $f(z_j)$, respectively, and the covariance matrix of $Y = [y_1, \ldots, y_N]$ and $f_Z = [f(z_1), \cdots, f(z_n)]$ is $k_{XZ} = (k(x_i, z_j))_{i=1,\ldots,N, j=1,\ldots,n}$. In summary, the simultaneous distribution of Y and f_Z is

$$\begin{bmatrix} Y \\ f_Z \end{bmatrix} \sim N \left(\begin{bmatrix} m_X \\ m_Z \end{bmatrix}, \begin{bmatrix} k_{XX} + \sigma^2 I & k_{XZ} \\ k_{ZX} & k_{ZZ} \end{bmatrix} \right).$$

In the following, we show that the posterior probability of the function $f(\cdot)$ given the value of Y is still a Gaussian process. To this end, we use the following proposition.

Proposition 61 *Suppose that the simultaneous distribution of random variables $a \in \mathbb{R}^N, b \in \mathbb{R}^n$ can be expressed by*

$$\begin{bmatrix} a \\ b \end{bmatrix} \sim N \left(\begin{bmatrix} \mu_a \\ \mu_b \end{bmatrix}, \begin{bmatrix} A & C \\ C^\top & B \end{bmatrix} \right)$$

where μ_a, μ_b are the expectations, $A \in \mathbb{R}^{N \times N}$, $B \in \mathbb{R}^{n \times n}$ are the covariance matrices (A: positive definite; B: nonnegative definite), and $C \in \mathbb{R}^{N \times n}$ is the covariance matrix between them. Then, the conditional probability of b given a is

$$b|a \sim N(\mu_b + C^\top A^{-1}(a - \mu_a), B - C^\top A^{-1}C). \tag{6.2}$$

Proof: Consult Lauritzen, "Graphical Models" [20] p256.

Hence, from Proposition 61, the posterior distribution of $f_Z \in \mathbb{R}^n$ under $Y \in \mathbb{R}^N$ is $N(\mu', \Sigma')$, where

$$\mu' := m_Z + k_{ZX}(k_{XX} + \sigma^2 I)^{-1}(Y - m_X) \in \mathbb{R}^n$$

and

$$\Sigma' := k_{ZZ} - k_{ZX}(k_{XX} + \sigma^2 I)^{-1}k_{XZ} \in \mathbb{R}^{n \times n}.$$

If we set $n = 1$ and $z_1 = x$, then the distribution of $f(x)$ becomes

$$m'(x) := m(x) + k_{xX}(k_{XX} + \sigma^2 I)^{-1}(Y - m_X) \tag{6.3}$$

$$k'(x, x) := k(x, x) - k_{xX}(k_{XX} + \sigma^2 I)^{-1}k_{Xx}. \tag{6.4}$$

We summarize the discussion as follows.

Proposition 62 *Suppose that the prior distribution of $f(\cdot)$ is a Gaussian process (m, k). If we obtain $x_1, \ldots, x_N, y_1, \ldots, y_N$ according to (6.1), the posterior probability of $f(\cdot)$ is a Gaussian process (m', k'), where m', k' are given by (6.3) and (6.4), respectively.*

In the actual calculation, it takes $O(N^3)$ time to calculate $(K + \sigma^2 I)^{-1}$. To complete the whole process in $O(N^3/3)$, we use the following method. By Cholesky decomposition, we obtain an $L \in \mathbb{R}^{N \times N}$ such that

$$LL^\top = k_{XX} + \sigma^2 I$$

which can be completed in $O(N^3/3)$ time. Then, let the solutions of $L\gamma = k_{Xx}$, $L\beta = y - m(x)$, and $L^\top \alpha = \beta$ be $\gamma \in \mathbb{R}^N$, $\beta \in \mathbb{R}^N$, and $\alpha \in \mathbb{R}^N$, respectively. Since L is a lower triangular matrix, these calculations take at most $O(N^2)$ time. Additionally, we have

$$(k_{XX} + \sigma^2 I)^{-1}(Y - m_X) = (LL^\top)^{-1}L\beta = (LL^\top)^{-1}LL^\top \alpha = \alpha$$

and

$$k_{xX}(k_{XX} + \sigma^2 I)^{-1}k_{Xx} = (L\gamma)^\top (LL^\top)^{-1}L\gamma = \gamma^\top \gamma.$$

Finally, from α, β, γ, we have

$$m'(x) = m(x) + k_{xX}\alpha$$

and

$$k'(x, x) = k(x, x) - \gamma^\top\gamma.$$

We can write the calculations of $m(x)$, $k(x, x)$ in forms that are completed in $O(N^3)$ and $O(N^3/3)$ time in the R language as follows.

```
gp.1=function(x.pred){
  h=array(dim=n); for(i in 1:n)h[i]=k(x.pred,x[i])
  R=solve(K+sigma.2*diag(n))                    ## O(n^3) Computation
  mm=mu(x.pred)+t(h)%*%R%*%(y-mu(x))
  ss=k(x.pred,x.pred)-t(h)%*%R%*%h
  return(list(mm=mm,ss=ss))
}
gp.2=function(x.pred){
  h=array(dim=n); for(i in 1:n)h[i]=k(x.pred,x[i])
  L=chol(K+sigma.2*diag(n))                      ## O(n^3/3) Computation
  alpha=solve(L,solve(t(L),y-mu(x)))             ## O(n^2) Computation
  mm=mu(x.pred)+sum(t(h)*alpha)
  gamma=solve(t(L),h)                            ## O(n^2) Computation
  ss=k(x.pred,x.pred)-sum(gamma^2)
  return(list(mm=mm,ss=ss))
}
```

Example 85 For comparison purposes, we executed the functions gp.1 and gp.2. We can see the difference achieved by Cholesky decomposition, which reduced the computational complexity.

```
sigma.2=0.2
k=function(x,y)exp(-(x-y)^2/2/sigma.2)   # Covariance function
mu=function(x) x                         # Mean function
n=1000; x=runif(n)*6-3; y=sin(x/2)+rnorm(n)   # Data Generation
K=array(dim=c(n,n)); for(i in 1:n)for(j in 1:n)K[i,j]=k(x[i],x[j])
## Measure Execution Time
library(tictoc)
tic(); gp.1(0); toc()
tic(); gp.2(0); toc()
# The 3 sigma width around the average
u.seq=seq(-3,3,0.1); v.seq=NULL; w.seq=NULL;
for(u in u.seq){res=gp.1(u); v.seq=c(v.seq,res$mm); w.seq=c(w.seq,
    sqrt(res$ss))}
plot(u.seq,v.seq,xlim=c(-3,3),ylim=c(-3,3),type="l")
lines(u.seq,v.seq+3*w.seq,col="blue"); lines(u.seq,v.seq-3*w.seq,col
    ="blue")
points(x,y)
## Five times, changing the samples
plot(0,xlim=c(-3,3),ylim=c(-3,3),type="n")
n=100
```

```
for(h in 1:5){
  x=runif(n)*6-3; y=sin(pi*x/2)+rnorm(n)
  sigma2=0.2
  K=array(dim=c(n,n)); for(i in 1:n)for(j in 1:n)K[i,j]=k(x[i],x[j])
  u.seq=seq(-3,3,0.1); v.seq=NULL
  for(u in u.seq){res=gp.1(u); v.seq=c(v.seq,res$mm)}
  lines(u.seq,v.seq,col=h+1)
}
```

If we compare the equation

$$m'(x) := k_{xX}(k_{XX} + \sigma^2 I)^{-1} Y$$

obtained by substituting $m_X = m(x) = 0$ into the average formula of the Gaussian process (6.3) with the equation

$$k_{x,X}\hat{\alpha} = k_{xX}(K + \lambda I)^{-1} Y$$

obtained by multiplying the kernel ridge regression formula (4.6) by $k_{x,x}$ from the left, we observe that the former is a specific case of the latter when setting $\lambda = \sigma^2$ (Fig. 6.1).

6.2 Classification

We consider the classification problem next. We assume that the random variable Y takes the value $Y = \pm 1$ and that its conditional probability given $x \in E$ is

$$P(Y = 1|x) = \frac{1}{1 + \exp(-f(x))} \tag{6.5}$$

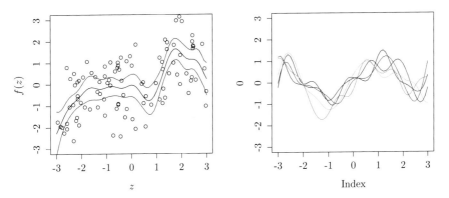

Fig. 6.1 We show the range of 3 σ above and below the average (left) and executed different samples five times (right)

where the Gaussian process $f : \Omega \times E \to \mathbb{R}$ is used. We wish to estimate f from the actual $x_1, \ldots, x_N \in \mathbb{R}^p$ (row vector) and $y_1, \ldots, y_N \in \{-1, 1\}$. To maximize the likelihood, we minimize the negative log-likelihood

$$\sum_{i=1}^{N} \log[1 + \exp\{-y_i f(x_i)\}].$$

If we set $f_X = [f_1, \ldots, f_N]^\top = [f(x_1), \ldots, f(x_N)]^\top \in \mathbb{R}^N$, $v_i := e^{-y_i f_i}$, and $l(f_X) := \sum_{i=1}^{N} \log(1 + v_i)$, then we have

$$\frac{\partial v_i}{\partial f_i} = -y_i v_i \, , \quad \frac{\partial l(f_X)}{\partial f_i} = -\frac{y_i v_i}{1 + v_i} \, , \quad \frac{\partial^2 l(f_X)}{\partial f_i^2} = \frac{v_i}{(1 + v_i)^2}$$

where we use $y_i^2 = 1$. Given an initial value, we wait for the Newton-Raphson update $f_X \leftarrow f_X - (\nabla^2 l(f_X))^{-1} \nabla l(f_X)$ to converge. The update formula is

$$f_X \leftarrow f_X + W^{-1} u$$

where $u = \left(\dfrac{y_i v_i}{1 + v_i} \right)_{i=1,\ldots,N}$ and $W = \text{diag}\left(\dfrac{v_i}{(1 + v_i)^2} \right)_{i=1,\ldots,N}$. In other words, we repeat the following two steps:

1. Obtain v, u, and W from f_X
2. Calculate $f_X + W^{-1} u$ and substitute it into f_X

for $v := [v_1, \ldots, v_N]^\top \in \mathbb{R}^N$.

Next, we consider maximizing the likelihood $\displaystyle\prod_{i=1}^{N} \frac{1}{1 + \exp\{-y_i f(x_i)\}}$ multiplied by the prior distribution of f_X, i.e., finding the solution with the maximum posterior probability. Here, the mean is often set to 0 as the prior probability of f in the formulation of (6.5). Suppose first that the prior probability of $f_X \in \mathbb{R}^N$ is

$$\frac{1}{\sqrt{(2\pi)^N \det k_{XX}}} \exp\{-\frac{f_X^\top k_{XX}^{-1} f_X}{2}\}$$

where k_{XX} is the Gram matrix $(k(x_i, x_j))_{i,j=1,\ldots,N}$. If we set

$$L(f_X) = l(f_X) + \frac{1}{2} f_X^\top k_{XX}^{-1} f_X + \frac{1}{2} \log \det k_{XX} + \frac{N}{2} \log 2\pi \qquad (6.6)$$

then we have

$$\nabla L(f_X) = \nabla l(f_X) + k_{XX}^{-1} f_X = -u + k_{XX}^{-1} f_X \qquad (6.7)$$

and

$$\nabla^2 L(f_X) = \nabla^2 l(f_X) + k_{XX}^{-1} = W + k_{XX}^{-1}. \qquad (6.8)$$

Thus, we may express the update formula as

$$f_X \leftarrow f_X + (W + k_{XX}^{-1})^{-1}(u - k_{XX}^{-1}f_X)$$
$$= (W + k_{XX}^{-1})^{-1}\{(W + k_{XX}^{-1})f_X - k_{XX}^{-1}f_X + u\}$$
$$= (W + k_{XX}^{-1})^{-1}(Wf_X + u).$$

However, since the size of f_X is the number of samples N, it takes an enormous amount of time to calculate the inverse matrix. We try to improve the efficiency of this process as follows. Utilizing the Woodbury-Sherman-Morrison formula

$$(A + UWV^{\top})^{-1} = A^{-1} - A^{-1}U(W^{-1} + V^{\top}A^{-1}U)^{-1}V^{\top}A^{-1} \qquad (6.9)$$

with $A \in \mathbb{R}^{n \times n}$ (nonsingular), $W \in \mathbb{R}^{m \times m}$, and $U, V \in \mathbb{R}^{n \times m}$, if we set $A = k_{XX}^{-1}$ and $U = V = I$, we obtain

$$(W + k_{XX}^{-1})^{-1}$$
$$= k_{XX} - k_{XX}(W^{-1} + k_{XX})^{-1}k_{XX}$$
$$= k_{XX} - k_{XX}W^{1/2}(I + W^{1/2}k_{XX}W^{1/2})^{-1}W^{1/2}k_{XX}. \qquad (6.10)$$

Thus, we can obtain an L such that $I + W^{1/2}k_{XX}W^{1/2} = LL^{\top}$ (Cholesky decomposition) in $O(N^3/3)$ time. Letting $\gamma := Wf_X + u$, we find a β such that $L\beta = W^{1/2}k_{XX}\gamma$ and an α such that $L^{\top}W^{-1/2}\alpha = \beta$ in $O(N^2)$ time, and we substitute $k_{XX}(\gamma - \alpha)$ into f_X. We repeat this procedure until convergence is achieved. In fact, we have the following equation:

$$LL^{\top}W^{-1/2}\alpha = L\beta = W^{1/2}k_{XX}\gamma$$

$$k_{XX}(\gamma - \alpha)$$
$$= k_{XX}\{\gamma - W^{1/2}(LL^{\top})^{-1}W^{1/2}k_{XX}\gamma\} = \{k_{XX} - k_{XX}W^{1/2}(LL^{\top})^{-1}W^{1/2}k_{XX}\}\gamma$$
$$= \{k_{XX} - k_{XX}W^{1/2}(I + W^{1/2}k_{XX}W^{1/2})^{-1}W^{1/2}k_{XX}\}\gamma = (W + k_{XX}^{-1})^{-1}(Wf + u)$$

where the last equality is due to (6.10).

Example 86 By using the first $N = 100$ of the 150 Iris data (the first 50 points and the next 50 points are Setosa and Versicolor data, respectively), we found the $f_X = [f_1, \ldots, f_N]$ with the maximum posterior probability. The output showed that $f_1, \ldots, f_{50}$ and $f_{51}, \ldots, f_{100}$ were positive and negative, respectively.

```
1  ## Iris Data
2  df=iris
3  x=df[1:100,1:4]
4  y=c(rep(1,50),rep(-1,50))
5  n=length(y)
6  ## Compute Kernel values for the four covariates
7  k=function(x,y) exp(sum(-(x-y)^2)/2)
8  K=matrix(0,n,n)
9  for(i in 1:n)for(j in 1:n)K[i,j]=k(x[i,],x[j,])
```

```
10  eps=0.00001
11  f=rep(0,n)
12  g=rep(0.1,n)
13  while(sum((f-g)^2)>eps){
14    g=f        ## Save the data before update for comparison
15    v=exp(-y*f)
16    u=y*v/(1+v)
17    w=as.vector(v/(1+v)^2)
18    W=diag(w); W.p=diag(w^0.5); W.m=diag(w^(-0.5))
19    L=chol(diag(n)+W.p%*%K%*%W.p)
20    L=t(L)    ## The chol function in R outputs the transposed matrix
21    gamma=W%*%f+u
22    beta=solve(L,W.p%*%K%*%gamma)
23    alpha=solve(t(L)%*%W.m,beta)
24    f=K%*%(gamma-alpha)
25  }
26  as.vector(f)
```

> as.vector(f)
 [1] 2.901760 2.666188 2.736000 2.596215 2.888826 2.422990 2.712837 2.896583
 [9] 2.263840 2.722794 2.675787 2.804277 2.629917 2.129059 1.994737 1.725577
[17] 2.502404 2.894768 2.211715 2.757889 2.580703 2.788434 2.450147 2.598253
[25] 2.493633 2.589272 2.799560 2.854338 2.858034 2.682198 2.663180 2.652952
[33] 2.409809 2.157029 2.738196 2.777507 2.605493 2.848624 2.342636 2.882683
[41] 2.887406 1.561917 2.454169 2.649399 2.407117 2.633906 2.727124 2.673216
[49] 2.749571 2.884288 -1.870442 -2.537382 -1.932737 -2.531886 -2.579367 -2.785015
[57] -2.381783 -1.467521 -2.486519 -2.356974 -1.600772 -2.811324 -2.381371 -2.734406
[65] -2.330770 -2.341664 -2.614817 -2.748088 -2.372972 -2.628800 -2.251908 -2.789266
[73] -2.345693 -2.704238 -2.712489 -2.502956 -2.162480 -2.015710 -2.840261 -2.194749
[81] -2.474090 -2.342643 -2.755051 -2.189084 -2.415174 -2.382286 -2.275106 -2.496755
[89] -2.695880 -2.664357 -2.648632 -2.771846 -2.807510 -1.546826 -2.814728 -2.756341
[97] -2.834644 -2.849801 -1.182284 -2.845863

To execute classification for a new value x by using the estimated $\hat{f} \in \mathbb{R}^N$, we perform the following steps. Similar to the regression case, if we apply Proposition 61 to

$$\begin{bmatrix} f_X \\ f(x) \end{bmatrix} \sim N \left(\begin{bmatrix} 0 \\ 0 \end{bmatrix}, \begin{bmatrix} k_{XX} & k_{Xx} \\ k_{xX} & k_{xx} \end{bmatrix} \right)$$

then we obtain

$$f(x)|f_X \sim N(m'(x), k'(x,x)) \qquad (6.11)$$

where

$$m'(x) = k_{xX} k_{XX}^{-1} f_X$$

and

$$k'(x,x) = k_{xx} - k_{xX} k_{XX}^{-1} k_{Xx}.$$

If we observe $Y \in \{-1, 1\}^N$, we obtain the estimate $\hat{f}$ with the Newton-Raphson method and calculate $\hat{W}$. We consider the Laplace approximation of $f_X|Y$, i.e., we approximate the Gaussian distribution as follows (Rasmussen-Williams [25]):

$$f_X|Y \sim N(\hat{f}, (\hat{W} + k_{XX}^{-1})^{-1}). \tag{6.12}$$

That is, the covariance matrix is the inverse of $\hat{W} + k_{XX}^{-1}$, which is the Hessian $\nabla^2 L(\hat{f})$ of (6.8). Then, the variations in (6.11)(6.12) are independent, and $f(x|Y) = N(m_*, k_*)$. Note that we can calculate the posterior probability for each $x \in E$:

$$m_* = k_{xX} k_{XX}^{-1} \hat{f} \tag{6.13}$$

$$
\begin{aligned}
k_* &= k_{xx} - k_{xX} k_{XX}^{-1} k_{Xx} + k_{xX} k_{XX}^{-1} (\hat{W} + k_{XX}^{-1})^{-1} k_{XX}^{-1} k_{Xx} \\
&= k_{xx} - k_{xX} (\hat{W}^{-1} + k_{XX})^{-1} k_{Xx}
\end{aligned}
$$

where the last transformation is due to $A = k_{XX}$ and $W = \hat{W}^{-1}$ in (6.9), and U, V are the unit matrices. Thus, we can compute the expectation (prediction value) w.r.t. $f(x)|Y \sim N(m_*, k_*)$ in the sigmoid function

$$P(Y = 1|x) = \frac{1}{1 + \exp(-f(x))} \quad :$$

$$\int_E \frac{1}{1 + \exp(-z)} \frac{1}{\sqrt{2\pi k_*}} \exp[-\frac{1}{2k_*}\{z - m_*\}^2] dz. \tag{6.14}$$

To implement this step, we only need to compute $\hat{u}$ from $\hat{f}$. Since (6.7) is zero when the updates converge, from (6.13), we have that

$$m_* = k_{xX} \hat{u}$$

and

$$
\begin{aligned}
(k_{XX} + W^{-1})^{-1} &= W^{1/2} W^{-1/2} (k_{XX} + W^{-1})^{-1} W^{-1/2} W^{1/2} \\
&= W^{1/2} (I + W^{1/2} k_{XX} W^{1/2})^{-1} W^{1/2}.
\end{aligned}
$$

Hence, we compute $\alpha \in \mathbb{R}^N$ such that $I + \hat{W}^{1/2} k_{XX} \hat{W}^{1/2} = LL^\top$ (Cholesky decomposition) and $L\alpha = \hat{W}^{1/2} k_{Xx}$ in $O(N^3/3)$ time. Then, we have

$$k_* = k_{xx} - \alpha^\top \alpha$$

since we have

$$k_{xX} W^{1/2} (LL^\top)^{-1} W^{1/2} k_{Xx} = k_{xX} W^{1/2} (L^{-1})^\top L^{-1} W^{1/2} k_{Xx} = \alpha^\top \alpha.$$

We can describe the procedure for finding the value of (6.14) in the R language as follows. We assume that the procedure starts immediately after the procedure of Example 86 completes.

```
pred=function(z){
    kk=array(0,dim=n); for (i in 1:n)kk[i]=k(z,x[i,])
    mu=sum(kk*as.vector(u))          ## Mean
    alpha=solve(L,W.p%*%kk); sigma2=k(z,z)-sum(alpha^2)      ## Variance
    m=1000; b=rnorm(m,mu,sigma2); pi=sum((1+exp(-b))^(-1))/m    ##
    Prediction
    return(pi)
}
```

Example 87 Immediately after processing Example 86, we entered numerical values for the four covariates of Iris into the function pred and calculated the probability of them being Setosa values (1 minus the probability of them being Versicolor values). When we input the average values of the covariates for Setosa and Versicolor, we observed that the probabilities were close to 1 and 0, respectively.

```
> z=array(0,dim=4)
> for(j in 1:4)z[j]=mean(x[1:50,j])
> pred(z)
[1] 0.9455452
> for(j in 1:4)z[j]=mean(x[51:100,j])
> pred(z)
[1] 0.05344474
```

6.3   Gaussian Processes with Inducing Variables

A Gaussian process generally involves $O(N^3)$ calculations. To avoid such an inconvenience, we choose $Z := [z_1, \cdots, z_M] \in E^M$ and approximate the generation process

$$f_X \sim N(m_X, k_{XX})$$

$$f(x)|f_X \sim N(m(x) + k_{xX}k_{XX}^{-1}(f_X - m_X), k(x,x) - k_{xX}k_{XX}^{-1}k_{Xx})$$

$$y|f(x) \sim N(f(x), \sigma^2)$$

by

$$f_Z \sim N(m_Z, k_{ZZ}) \tag{6.15}$$

$$f(x)|f_Z \sim N(m(x) + k_{xZ}k_{ZZ}^{-1}(f_Z - m_Z), k(x,x) - k_{xZ}k_{ZZ}^{-1}k_{Zx}) \tag{6.16}$$

$$y|f(x) \sim N(f(x), \sigma^2), \tag{6.17}$$

where $\quad m_Z = (m(z_1), \cdots, m(z_M))$, $\quad k_{ZZ} = (k(z_i, z_j))_{i,j=1,\cdots,M}$, $\quad$ and $\quad k_{xZ} = [k(x, z_1), \cdots, k(x, z_M)]$ (row vector).

Under the following assumption, we obtain Proposition 63.

Assumption 1 Each occurrence of (6.16) is independent for $x = x_1, \cdots, x_N$.

Proposition 63 Let $\Lambda \in \mathbb{R}^{N \times N}$ be a diagonal matrix whose elements are $\lambda(x_i)$, $i = 1, \cdots, N$. Under the generation process outlined in (6.15), (6.16), (6.17) and Assumption 1, we have

$$f_Z | Y \sim N(\mu_{f_Z|Y}, \Sigma_{f_Z|Y})$$

$$\mu_{f_Z|Y} = m_Z + k_{ZZ} Q^{-1} k_{ZX} (\Lambda + \sigma^2 I_N)^{-1} (Y - m_X) \tag{6.18}$$

and

$$\Sigma_{f_Z|Y} = k_{ZZ} Q^{-1} k_{ZZ}, \tag{6.19}$$

where

$$Q := k_{ZZ} + k_{ZX} (\Lambda + \sigma^2 I_N)^{-1} k_{XZ} \in \mathbb{R}^{M \times M} \tag{6.20}$$

with $\lambda(x) := k(x, x) - k_{xZ} k_{ZZ}^{-1} k_{Zx}$.

Proof: From (6.16) and Assumption 1,

$$f_X | f_Z \sim N(m_X + k_{XZ} k_{ZZ}^{-1} (f_Z - m_Z), \Lambda)$$

for $f_X := [f(x_1), \cdots, f(x_N)]$. Moreover, the expectations of Y and f_X are equal, and only the variance σ^2 is different, so we have

$$Y | f_Z \sim N(m_X + k_{XZ} k_{ZZ}^{-1} (f_Z - m_Z), \Lambda + \sigma^2 I_N).$$

Thus, the simultaneous distribution of f_Z, Y is

$$\text{(the exponents of } p(f_Z) \text{ and } p(Y|f_Z))$$
$$= -\frac{1}{2} (f_Z - m_Z)^\top k_{ZZ}^{-1} (f_Z - m_Z)$$
$$- \frac{1}{2} \{ Y - (m_X + k_{ZX} k_{ZZ}^{-1} (f_Z - m_Z)) \}^\top (\Lambda + \sigma^2 I_N)^{-1}$$
$$\cdot \{ Y - (m_X + k_{ZX} k_{ZZ}^{-1} (f_Z - m_Z)) \}. \tag{6.21}$$

If we differentiate (6.21) by f_Z, setting $a = f_Z - m_Z$ and $b = Y - m_X$ yields

$$-k_{ZZ}^{-1} a + k_{ZZ}^{-1} k_{ZX} (\Lambda + \sigma^2 I_N)^{-1} (b - k_{ZX} k_{ZZ}^{-1} a)$$
$$= k_{ZZ}^{-1} k_{ZX} (\Lambda + \sigma^2 I_N)^{-1} b - k_{ZZ}^{-1} \{ k_{ZZ} + k_{ZX} (\Lambda + \sigma^2 I_N)) k_{XZ} \} k_{ZZ}^{-1} a$$
$$= k_{ZZ}^{-1} k_{ZX} (\Lambda + \sigma^2 I_N)^{-1} b - k_{ZZ}^{-1} Q k_{ZZ}^{-1} a$$

$$\begin{aligned}
&= k_{ZZ}^{-1} Q k_{ZZ}^{-1} \{ k_{ZZ} Q^{-1} k_{ZX} (\Lambda + \sigma^2 I_N)^{-1} b - a \} \\
&= -\Sigma_{f_Z|Y}^{-1} (f_Z - \mu_{f_Z|Y}).
\end{aligned} \tag{6.22}$$

Therefore, the terms w.r.t. f_Z in (6.21) are only

$$-\frac{1}{2}(f_Z - \mu_{f_Z|Y})^\top \Sigma_{f_Z|Y}^{-1} (f_Z - \mu_{f_Z|Y}), \tag{6.23}$$

and we obtain the proposition. $\square$

Proposition 64 *Under the generation process outlined in (6.15), (6.16), (6.17) and Assumption 1, we have*

$$Y \sim N(\mu_Y, \Sigma_Y)$$

with

$$\mu_Y := m_X \tag{6.24}$$

and

$$\Sigma_Y := \Lambda + \sigma^2 I_N + k_{XZ} k_{ZZ}^{-1} k_{XZ}. \tag{6.25}$$

Proof: Since the expectation μ_Y of Y is m_X, we obtain the covariance matrix Σ_Y. Let $a := f_Z - m_Z$ and $b := Y - m_X$. Then, the exponents (6.21) and (6.23) of $p(Y, f_Z)$ and $p(f_Z|Y)$ are, respectively,

$$-\frac{1}{2} a^\top k_{ZZ}^{-1} a - \frac{1}{2}(b - k_{ZX} k_{ZZ}^{-1} a)^\top (\Lambda + \sigma^2 I)^{-1} (b - k_{ZX} k_{ZZ}^{-1} a) \tag{6.26}$$

and

$$-\frac{1}{2}(a - k_{ZZ} Q^{-1} k_{ZX}(\Lambda + \sigma^2 I_N)^{-1} b)^\top (k_{ZZ} Q^{-1} k_{ZZ})^{-1} (a - k_{ZZ} Q^{-1} k_{ZX}(\Lambda + \sigma^2 I_N)^{-1} b). \tag{6.27}$$

From (6.20), we have

$$-\frac{1}{2} a^\top k_{ZZ}^{-1} a - \frac{1}{2}(k_{ZX} k_{ZZ}^{-1} a)^\top (\Lambda + \sigma^2 I)^{-1} k_{ZX} k_{ZZ}^{-1} a = -\frac{1}{2} a^\top k_{ZZ}^{-1} Q k_{ZZ}^{-1} a.$$

From $p(Y, f_2) = p(f_2|Y) p(Y)$, the difference between (6.26) and (6.27) is the exponent of $p(Y)$, which is

$$-\frac{1}{2} b^\top (\Lambda + \sigma^2 I)^{-1} b + \frac{1}{2} b^\top (\Lambda + \sigma^2 I_N)^{-1} k_{XZ} Q^{-1} k_{ZX}(\Lambda + \sigma^2 I_N)^{-1} b$$

where we may set $a = 0$ because no terms will remain w.r.t. a. Furthermore, if we set $A = \Lambda + \sigma^2 I_N$, $U = k_{XZ}$, $V = k_{ZX}$, and $W = k_{ZZ}^{-1}$ in the Woodbury-Sherman-Morrison formula (6.9), then we have

$$-\frac{1}{2}b^{\top}(\Lambda + \sigma^2 I_N + k_{XZ}k_{ZZ}^{-1}k_{XZ})^{-1}b$$

and obtain (6.25).   □

Proposition 65 *Under the generation process outlined in (6.15), (6.16), (6.17) and Assumption 1, for each $x \in E$, we have*

$$f(x)|Y \sim N(\mu(x), \sigma^2(x))$$

$$\mu(x) := m(x) + k_{xZ}k_{ZZ}^{-1}(\mu_{f_Z|Y} - m_Z) = m(x) + k_{xZ}Q^{-1}k_{ZX}(\Lambda + \sigma^2 I_N)^{-1}(Y - m_X) \tag{6.28}$$

$$\sigma^2(x) := k(x, x) - k_{xZ}(K_{ZZ}^{-1} - Q^{-1})k_{Zx}.$$

Proof: First, we note that $Y \to f_Z \to f(x)$ forms a Markov chain in this order. In the following, we consider the distribution of $f(x)|Y$ instead of $f(x)|f_Z$, i.e., the distribution of $f(x)|f_Z$ and $f_Z|Y$. In (6.16), the term with a mean of $k_{xZ}k_{ZZ}^{-1}(f_Z - m_Z)$ becomes $k_{xZ}k_{ZZ}^{-1}(\mu_{f_Z|Y} - m_Z)$ when averaged over $f_Z|Y$. Thus, we obtain (6.28). Moreover, if we take the variance of that term with respect to $f_Z|Y$, we obtain the same value as the variance of $k_{xZ}k_{ZZ}^{-1}(f_Z - \mu_{f_Z|Y})$, so we have

$$\mathbb{E}[k_{xZ}k_{ZZ}^{-1}(f_Z - \mu_{f_Z|Y})(f_Z - \mu_{f_Z|Y})^{\top}k_{ZZ}^{-1}k_{Zx}] = k_{xZ}k_{ZZ}^{-1}\Sigma_{f_Z|Y}k_{ZZ}^{-1}k_{Zx} = k_{xZ}Q^{-1}k_{Zx} \tag{6.29}$$

where f_Z varies with the given Y. Furthermore, from (6.16), since the variance $\lambda(x) = k(x, x) - k_{xZ}k_{ZZ}^{-1}k_{Zx}$ of $f(x)|f_Z$ is independent of f_Z, we can write the variance of $f(x)|Y$ as the sum of the variance $\lambda(x)$ of $f(x)|f_Z$ and (6.29). In other words, we have $\sigma^2(x) = \lambda(x) + k_{xZ}Q^{-1}k_{Zx}$.   □

In a case when the inducing variable method is employed, the calculations of k_{ZZ}, k_{xZ} take $O(M^2)$ and $O(M)$, respectively, the calculation of Λ takes $O(N)$, and the calculations of Q and Q^{-1} take $O(NM^2)$ and $O(M^3)$, respectively. The multiplication process is also completed in $O(NM^2)$. On the other hand, without the inducing variable method, it takes $O(N^3)$ of computational time. In the inducing variable method, we do not use the matrix $K_{XX} \in \mathbb{R}^{N \times N}$.

We can randomly select the inducing points $z_1, \cdots, z_M$ from $x_1, \cdots, x_N$ or via K-means clustering.

Example 88 Based on the above discussion, we constructed the function gp.ind by using the inducing variable method and compared its performance with that of gp.1, which does not use the inducing variable method.

```
1  sigma.2=0.05         # should be estimated
2  k=function(x,y)exp(-(x-y)^2/2/sigma.2)    # Covariance function
3  mu=function(x) x                          # Mean function
4  n=200; x=runif(n)*6-3; y=sin(x/2)+rnorm(n)   # Data Generation
5  eps=10^(-6)
6
7  m=100
8  index=sample(1:n, m, replace=FALSE)
9  z=x[index]
10 m.x=0
11 m.z=0
12 K.zz=array(dim=c(m,m)); for(i in 1:m)for(j in 1:m)K.zz[i,j]=k(z[i],z
       [j])
13 K.xz=array(dim=c(n,m)); for(i in 1:n)for(j in 1:m)K.xz[i,j]=k(x[i],z
       [j])
14 K.zz.inv=solve(K.zz+diag(rep(10^eps,m)))
15 lambda=array(dim=n)
16 for(i in 1:n)lambda[i]=k(x[i],x[i])-K.xz[i,1:m]%*%K.zz.inv%*%K.xz[i
       ,1:m]
17 Lambda.0.inv=diag(1/(lambda+sigma.2))
18 Q=K.zz+t(K.xz)%*%Lambda.0.inv%*%K.xz    ## Computation of Q does not require
       O(n^3)
19 Q.inv=solve(Q+diag(rep(eps,m)))
20 muu=Q.inv%*%t(K.xz)%*%Lambda.0.inv%*%(y-m.x)
21 dif=K.zz.inv-Q.inv
22 K=array(dim=c(n,n)); for(i in 1:n)for(j in 1:n)K[i,j]=k(x[i],x[j])
23 R=solve(K+sigma.2*diag(n))                ## O(n^3) omputation  is required
24
25 gp.ind=function(x.pred){
26   h=array(dim=m); for(i in 1:m)h[i]=k(x.pred,z[i])
27   mm=mu(x.pred)+h%*%muu
28   ss=k(x.pred,x.pred)-h%*%dif%*%h
29   return(list(mm=mm,ss=ss))
30 }                                         ## Inducing Variable Method
31
32 gp.1=function(x.pred){
33   h=array(dim=n); for(i in 1:n)h[i]=k(x.pred,x[i])
34   mm=mu(x.pred)+t(h)%*%R%*%(y-mu(x))
35   ss=k(x.pred,x.pred)-t(h)%*%R%*%h
36   return(list(mm=mm,ss=ss))
37 }## W/O Inducing Variable Method
38
39 x.seq=seq(-2,2,0.1)
40 mmv=NULL; ssv=NULL
41 for(u in x.seq){
42   mmv=c(mmv,gp.ind(u)$mm)
43   ssv=c(ssv,gp.ind(u)$ss)
44 }
45 plot(0, xlim=c(-2,2),ylim=c(min(mmv),max(mmv)),type="n")
46 lines(x.seq,mmv,col="red")
47 lines(x.seq,mmv+3*sqrt(ssv),lty=3,col="red")
48 lines(x.seq,mmv-3*sqrt(ssv),lty=3,col="red")
49
```

```
50
51   x.seq=seq(-2,2,0.1)
52   mmv=NULL; ssv=NULL
53   for(u in x.seq){
54       mmv=c(mmv,gp.1(u)$mm)
55       ssv=c(ssv,gp.1(u)$ss)
56   }
57
58   lines(x.seq,mmv,col="blue")
59   lines(x.seq,mmv+3*sqrt(ssv),lty=3,col="blue")
60   lines(x.seq,mmv-3*sqrt(ssv),lty=3,col="blue")
61   points(x,y)
```

6.4 Karhunen-Lóeve Expansion

In this section, we continue to study the probability space $(\Omega, \mathcal{F}, P)$ and the map $f :$ $\Omega \times E \ni (\omega, x) \rightarrow f(\omega, x) \in H$. We assume that H is a general separable Hilbert space. In the following, we continue to denote $f(\omega, x)$ by $f(x)$ as a random variable for each $x \in E$. In particular, we assume that f is a mean-square continuous process, which is defined by

$$\lim_{n \to \infty} \mathbb{E}|f(\omega, x_n) - f(\omega, x)|^2 = 0 \tag{6.30}$$

for an arbitrary $\{x_n\}$ in E that converges to $x \in E$. We do not assume a Gaussian process, and we give the expectation at $x \in E$ and the covariance at $x, y \in E$ by

$$m(x) = \mathbb{E}f(\omega, x)$$

and

$$k(x, y) = \mathrm{Cov}(f(\omega, x), f(\omega, y)).$$

In Chap. 5, we obtained the expectation and covariance of $k(X, \cdot)$; in this section, however, $x, y \in E$ are not random, and the randomness of m, k is due to that of $f(\omega, \cdot)$.

In the following, we assume that E is compact (Fig. 6.2).

Proposition 66 *f is a mean-square continuous process if and only if m and k are continuous.*

Proof: See the Appendix at the end of this chapter.

In the following, we assume that $m \equiv 0$ to simplify the discussion. Since E is compact, we assume that the diameter of each E_i is less than or equal to $1/n$. However, each E_i is a metric space, and we define the diameter by the maximum distance among the elements in E_i. Thus, there exists a partition of E ($\cup_{i=1}^{M(n)} E_i = E$, $E_i \cap E_j = \phi, i \neq j$) and a number of partitions $M(n)$. Then, we define

Fig. 6.2 The red and blue
curves show the results
obtained by the inducing
variable and standard
Gaussian processes,
respectively

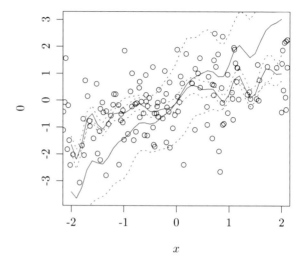

$$I_f(g; \{(E_i, x_i)\}_{1 \leq i \leq M(n)}) := \sum_{i=1}^{M(n)} f(\omega, x_i) \int_{E_i} g(y) d\mu(y)$$

for a pair of interior points $\{(E_i, x_i)\}_{1 \leq i \leq M(n)}$ and $g \in L^2(E, \mathcal{B}(E), \mu)$. Hence, we
have

$$\int_\Omega \{I_f(g; \{(E_i, x_i)\}_{1 \leq i \leq M(n)})\}^2 dP(\omega) \leq M(n) \sum_{i=1}^{M(n)} \int_\Omega \{f(\omega, x_i)\}^2 \{\int_{E_i} g(u) d\mu(u)\}^2 dP$$

$$= M(n) \sum_{i=1}^{M(n)} k(x_i, x_i) \int_{E_i} \{g(u)\}^2 d\mu(u) < \infty$$

and $I_f(g; \{(E_i, x_i)\}_{1 \leq i \leq M(n)}) \in L^2(\Omega, \mathcal{F}, P)$. Although this value is different
depending on the choices of the region decomposition and the points inside the
regions, the difference in I_f converges to zero as n goes to infinity. In fact, we have

$$\mathbb{E}\left[|I_f(g; \{(E_i, x_i)\}_{1 \leq i \leq M(n)}) - I_f(g; \{(E'_j, x'_j)\}_{1 \leq j \leq M(n')})|^2\right]$$

$$= \sum_{i=1}^{M(n)} \sum_{i'=1}^{M(n)} k(x_i, x_{i'}) \int_{E_i} g(u) d\mu(u) \int_{E_{i'}} g(v) d\mu(v)$$

$$+ \sum_{j=1}^{M(n')} \sum_{j'=1}^{M(n')} k(x_j, x_{j'}) \int_{E_j} g(u) d\mu(u) \int_{E_{j'}} g(v) d\mu(v)$$

$$-2 \sum_{i=1}^{M(n)} \sum_{j=1}^{M(n')} k(x_i, x_{j'}) \int_{E_i} g(u)d\mu(u) \int_{E_{j'}} g(v)d\mu(v).$$

Since k is uniformly continuous, each double sum on the right-hand side converges to

$$\int_E \int_E k(u, v)g(u)g(v)d\mu(u)d\mu(v).$$

Since the Cauchy sequence converges to zero, its convergence destination $I_f(\omega, g)$ is contained in $L^2(\omega, \mathcal{F}, P)$ regardless of the choice of $\{(E_i, x_i)\}_{1 \leq i \leq M(n)}$.

If the eigenvalues and eigenfunctions obtained from the integral operator $T_k \in B(L^2(E, \mathcal{B}(E), \mu))$,

$$T_k g(\cdot) = \int_E k(y, \cdot)g(y)d\mu(y) , \quad g \in L^2(E, \mathcal{B}(E), \mu),$$

are $\{\lambda_j\}_{j=1}^{\infty}$ and $\{e_j(\cdot)\}_{j=1}^{\infty}$, by Mercer's theorem, we can express the covariance function k as

$$k(x, y) = \sum_{j=1}^{\infty} \lambda_j e_j(x)e_j(y) \tag{6.31}$$

where the sum absolutely and uniformly converges on that support.

Then, we have the following claim.

Proposition 67 *If $\{f(\omega, x)\}_{x \in E}$ is a mean-square continuous process with a mean of zero, we have*

1. $\mathbb{E}[I_f(\omega, g)] = 0$
2. $\mathbb{E}[I_f(\omega, g)f(\omega, x)] = \int_E k(x, y)g(y)d\mu(y)$, $x \in E$
3. $\mathbb{E}[I_f(\omega, g)I_f(\omega, h)] = \int_E \int_E k(x, y)g(x)h(y)d\mu(x)d\mu(y)$

These properties hold for each $g, h \in L^2(E, \mathcal{F}, \mu)$, and in particular, we have that

$$\mathbb{E}[I_f(\omega, e_i)I_f(\omega, e_j)] = \delta_{i,j}\lambda_i. \tag{6.32}$$

Proof: For the proofs of the above three items, see the Appendix at the end of this chapter. We obtain (6.32) by substituting Mercer's theorem (6.31), $g = e_i$, and $h = e_j$ into the third item:

$$\mathbb{E}[I_f(\omega, e_i)I_f(\omega, e_j)] = \int_E \int_E \sum_{r=1}^{\infty} \lambda_r e_r(x)e_r(y)e_i(x)e_j(y)d\mu(x)d\mu(y).$$

□

Furthermore, we have the following theorem.

Proposition 68 (Karhunen-Lóeve [17, 18]) *Suppose that* $\{f(\omega, x)\}_{x \in E}$ *is a mean-square continuous process with a mean of zero. Then, we have*

$$\lim_{n \to \infty} \sup_{x \in E} \mathbb{E}|f(\omega, x) - f_n(\omega, x)|^2 = 0$$

for $f_n(\omega, x) := \sum_{j=1}^{n} I_f(\omega, e_j) e_j(x).$

Proof: From (6.32), we have

$$\mathbb{E}[f_n(\omega, x)^2] = \mathbb{E}[\{\sum_{j=1}^{n} I_f(\omega, e_j) e_j(x)\}^2]$$

$$= \sum_{i=1}^{n} \sum_{j=1}^{n} \mathbb{E}[I_f(\omega, e_i) I_f(\omega, e_j)] e_i(x) e_j(x) = \sum_{j=1}^{n} \lambda_j e_j^2(x).$$

Moreover, from (6.31) and the second item of Proposition 67, we have

$$\mathbb{E}[f_n(\omega, x) f(\omega, x)] = \mathbb{E}[\sum_{j=1}^{n} I_f(\omega, e_j) e_j(x) f(\omega, x)] = \sum_{j=1}^{n} e_j(x) \int_E k(x, y) e_j(y) d\mu(y)$$

$$= \sum_{j=1}^{n} \lambda_j e_j^2(x) \int_E e_j^2(y) d\mu(y) = \sum_{j=1}^{n} \lambda_j e_j^2(x)$$

which means that

$$\mathbb{E}|f_n(\omega, x) - f(\omega, x)|^2 = \mathbb{E}[f_n(\omega, x)^2] - 2\mathbb{E}[f_n(\omega, x) f(\omega, x)] + \mathbb{E}[f(\omega, x)^2]$$

$$= \sum_{j=1}^{n} \lambda_j e_j^2(x) - 2\sum_{j=1}^{n} \lambda_j e_j^2(x) + k(x, x) = k(x, x) - \sum_{j=1}^{n} \lambda_j e_j^2(x).$$

$\square$

In a general mean-square continuous process (without assuming a Gaussian process), the series expansion provided by Karhunen-Lóeve's theorem makes $I_f(\omega, e_j)/\sqrt{\lambda_j}$ a random variable with a mean of 0 and a variance of 1. Instead, if we assume a Gaussian process such that $f(x)$ $(x \in E)$ follows a Gaussian distribution, then we can write

$$f_n(x) = \sum_{j=1}^{n} z_j \sqrt{\lambda_j} e_j(x) \tag{6.33}$$

where z_j follows an independent standard Gaussian distribution.

Let $E := [0, 1]$ and $(\Omega, \mathcal{F}, P)$ be a probability space. Then, we call the map $\Omega \times E \ni (\omega, x) \mapsto f(\omega, x) \in \mathbb{R}$ that satisfies the following conditions a Brownian motion.

1. $f(\omega, 0) = 0$, $f(\omega, x) - f(\omega, y) \sim N(0, y - x)$, and $0 \le x < y$.
2. $f(\omega, x_2) - f(\omega, x_1), \dots, f(\omega, x_{n-1}) - f(\omega, x_n)$ are independent for any $n = 1, 2, \dots$ and $0 \le x_1 < x_2 < \dots < x_n$.
3. There exists an $\Omega \in \mathcal{F}$ with a probability of 1, and $E \ni x \mapsto f(\omega, x)$ is continuous for each $\omega \in \Omega$.

In this case, we have the following proposition.

Proposition 69 *The map $\Omega \times E \ni (\omega, x) \mapsto f(\omega, x) \in \mathbb{R}$ is a Brownian motion if and only if the following three conditions are satisfied simultaneously.*

1. *It is a Gaussian process.*
2. *The covariance function of $x, y \in E$ is given by $k(x, y) = \min(x, y)$.*
3. *$f(\omega, \cdot)$ is continuous with a probability of 1.*

Proof: The first two conditions in the definition imply the first condition in Proposition 69. Moreover, if $x < y$, then we have

$$\mathbb{E}[f(\omega, x) f(\omega, y)] = \mathbb{E}[f(\omega, x)^2] + \mathbb{E}[f(\omega, x)\{f(\omega, y) - f(\omega, x)\}] = x$$

which implies the second condition of Proposition 69. On the contrary, supposing that $m \equiv 0$ for simplicity, if we assume that the first two items of Proposition 69 hold, then because $k(x, x) = x$, when $x \le y \le z$, we have

$$\mathbb{E}[f(\omega, x)\{f(\omega, y) - f(\omega, z)\}] = k(x, y) - k(x, z) = x - x = 0$$

and

$$\mathbb{E}[f(\omega, y)\{f(\omega, y) - f(\omega, z)\}] = k(y, y) - k(y, z) = y - y = 0$$

which implies that

$$\mathbb{E}[\{f(\omega, x) - f(\omega, y)\}\{f(\omega, y) - f(\omega, z)\}] = 0.$$

Moreover, from $k(0, 0) = 0$, the variance of $f(\omega, 0)$ is zero, so we have $f(\omega, 0) = 0$. Furthermore, we have

$$\mathbb{E}[\{f(\omega, x) - f(\omega, y)\}^2] = k(x, x) - 2k(x, y) + k(y, y) = x - 2x + y = y - x.$$

$\square$

Example 89 (*Brownian Motion as a Gaussian Process*) For the integral operator (Example 58) on the covariance function of a Brownian motion $k(x, y) = \min(x, y)$ $(x, y \in E)$, its eigenvalues and eigenfunctions are (3.13) and (3.14), respectively. Utilizing these eigenvalues and eigenfunctions, we can expand $f(\omega, \cdot)$ as follows. In particular, from (6.33), we have

$$f_n(x) = \sum_{j=1}^{n} z_j(\omega) \sqrt{\lambda_j} e_j(x)$$

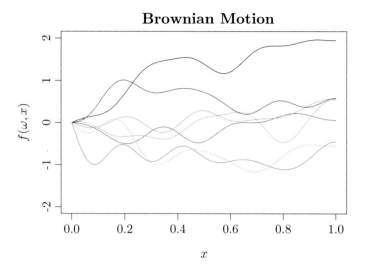

Fig. 6.3 We generated the sample paths of Brownian motions seven times. Each run involved a sum of up to 10 terms

for $z_j \sim N(0, 1)$. We generate the series $\{z_i(\omega)\}_{i=1}^n$ m times (Fig. 6.3; $n = 10$; $m = 7$). We execute it with the following code.

```
1   lambda=function(j) 4/((2*j-1)*pi)^2          ## EigenValue
2   ee=function(j,x) sqrt(2)*sin((2*j-1)*pi/2*x) ## Definition of Eigenfunction
3   n=10; m=7
4   f=function(z,x){                             ## Definition of Gaussian
        Process
5     n=length(z)
6     S=0; for(i in 1:n)S=S+z[i]*ee(i,x)*sqrt(lambda(i))
7     return(S)
8   }
9   plot(0,xlim=c(-3,3),ylim=c(-2,2),type="n",xlab="x",ylab="f(omega,x)
        ")
10  for(j in 1:m){
11    z=rnorm(n)
12    x.seq=seq(-3,3,0.001)
13    y.seq=NULL; for(x in x.seq)y.seq=c(y.seq,f(z,x))
14    lines(x.seq,y.seq,col=j)
15  }
16  title("Brown Motion")
```

We introduce the Matérn class, which is a class of kernels used in stochastic processes rather than the RKHS of machine learning. Such a kernel is $k(x, y) = \varphi(z)$ ($z := x - y$ in $x.y \in E$), where $\varphi(z)$ is

$$\varphi(z) := \frac{2^{1-\nu}}{\Gamma(\nu)\left(\frac{\sqrt{2\nu z}}{l}\right)} K_\nu\!\left(\frac{\sqrt{2\nu z}}{l}\right) \tag{6.34}$$

$\nu, l > 0$ are the parameters of the kernel, and K_ν is a variant Bessel function of the second kind.

$$K_\nu(x) := \frac{\pi}{2} \frac{I_{-\alpha}(x) - I_\alpha(x)}{\sin(\alpha x)}$$

$$I_\alpha(x) := \sum_{m=0}^{\infty} \frac{1}{m!\Gamma(m+\alpha+1)} \left(\frac{x}{2}\right)^{2m+\alpha}$$

In practice, we use (6.34), in which p is a positive integer and $\nu = p + 1/2$. In the 1-dimensional case, we have

$$\varphi_\nu(z) = \exp\left(-\frac{\sqrt{2\nu z}}{l}\right) \frac{\Gamma(p+1)}{\Gamma(2p+1)} \sum_{i=0}^{p} \frac{(p+i)!}{i!(p-i)!} \left(\frac{\sqrt{8\nu z}}{l}\right)^{p-i}. \tag{6.35}$$

For example, we express $\nu = 5/2, 3/2, 1/2$ as follows. In particular, we call the stochastic process with $\nu = 1/2$ the Ornstein-Uhlenbeck process.

$$\varphi_{5/2}(z) = \left(1 + \frac{\sqrt{5}z}{l} + \frac{5z^2}{3l^2}\right)\exp\left(-\frac{\sqrt{5}z}{l}\right)$$

$$\varphi_{3/2}(z) = \left(1 + \frac{\sqrt{3}z}{l}\right)\exp\left(-\frac{\sqrt{3}z}{l}\right)$$

$$\varphi_{1/2}(z) = \exp(-z/l)$$

For example, if we write this process in the R language, we have the following code.

```
matern=function(nu,l,r){
  p=nu-1/2
  S=0
  for(i in 0:p)S=S+gamma(p+i+1)/gamma(i+1)/gamma(p-i+1)*(sqrt(8*nu)*
    r/l)^(p-i)
  S=S*gamma(p+2)/gamma(2*p+1)*exp(-sqrt(2*nu)*r/l)
  return(S)
}
```

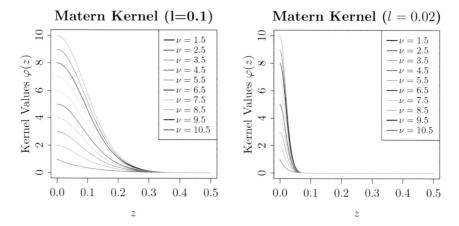

Fig. 6.4 The values of the Matérn kernel for $\nu = 1/2, 3/2, \ldots, m + 1/2$ (Example 90). $l = 0.1$ (left) and $l = 0.02$ (right)

Example 90 We present the Matérn kernel values for $l = 0.1, 0.02$ with $\nu = 1/2, 3/2, \ldots, m + 1/2$ (Fig. 6.4).

```
m=10
l=0.1
for(i in 1:1)curve(matern(i-1/2,l,x),0,0.5,ylim=c(0,10),col=i+1)
for(i in 2:m)curve(matern(i-1/2,l,x),0,0.5,ylim=c(0,10),ann=FALSE,
    add=TRUE,col=i+1)
legend("topright",legend=paste("nu=",(1:m)+0.5),lwd=2,col=1:m,)
title("Matern Kernel (l=1)")
```

In the case of the Matérn kernel and in general, we cannot analytically obtain the eigenvalues and eigenfunctions, as in the cases involving Gaussian kernels and Brownian motion. Even in those cases, if we assume a Gaussian process, then we can find $x_1, \ldots, x_n \in E$ to obtain its Gram matrix, which will be a covariance matrix. Thus, it is sufficient to generate n-variate random numbers that follow a Gaussian distribution. The above method is approximate, but it is very versatile.

Example 91 We display the Orstein-Uhlenbeck process ($\nu = 1.2$, top) and the Matérn process ($\nu = 3/2$, top) with $n = 100$ and $l = 0.1$ (Fig. 6.5).

```
rand.100=function(Sigma){
  L=t(chol(Sigma))        ## Cholesky decomposition of covariance matrix
  u=rnorm(100)
  y=as.vector(L%*%u)      ## Generate random numbers with zero mean and the
    covariance metrix
}

x = seq(0,1,length=100)
z = abs(outer(x,x,"-"))  # compute distance matrix, d_{ij} = |x_i - x_j|
```

```
9    l=0.1
10
11   Sigma_OU = exp(-z/l)        ## OU:  matern(1/2,l,z) is slow
12   y=rand.100(Sigma_OU)
13
14   plot(x,y,type="l",ylim=c(-3,3))
15   for(i in 1:5){
16     y = rand.100(Sigma_OU)
17     lines(x,y,col=i+1)
18   }
19   title("OU process (nu=1/2,l=0.1)")
20
21   Sigma_M=matern(3/2,l,z)    ## Matern
22   y = rand.100(Sigma_M)
23   plot(x,y,type="l",ylim=c(-3,3))
24   for(i in 1:5){
25     y = rand.100(Sigma_M)
26     lines(x,y,col=i+1)
27   }
28   title("Matern process (nu=3/2,l=0.1)")
```

6.5 Functional Data Analysis

Let $(\Omega, \mathcal{F}, P)$ and H be a probability space and a separable Hilbert space, respectively. Let $F : \Omega \to H$ be a measurable map, i.e., $\{h \in H| \|g - h\| < r\}$) is an element of $\mathcal{F}$ for each open set $(g \in H, r \in (0, \infty)$ in H. We call such an $F : \Omega \to H$ a random element of H. Intuitively, a random element is a random variable that takes a value in H. Thus far, we have assumed that $f : \Omega \times E \to \mathbb{R}$ is measurable at each $x \in E$ (stochastic process). This section addresses situations in which we do not assume such measurability. For simplicity, we write $F(\omega)$ as F, similar to the elements of H.

Although we do not go into details in this book, it is known that the following relationship holds between stochastic processes and random elements. It is only necessary to understand the close relationship between the two.

Proposition 70 (Hsing-Eubank [14])

1. If $f : \Omega \times E \to \mathbb{R}$ is measurable w.r.t. $\Omega \times E$ and $f(\omega, \cdot) \in H$, for $\omega \in \Omega$, then $f(\omega, \cdot)$ is a random element of H.
2. If $f(\cdot, x) \to \mathbb{R}$ is measurable for each $x \in E$ and $f(\omega, \cdot)$ is continuous for each $\omega \in \Omega$, then $f(\omega, \cdot)$ is a random element.
3. If $f : \Omega \times E \to \mathbb{R}$ is a (zero-mean) mean-square continuous process and its covariance function is k, then a random element of H exists such that the covariance operator is $H \ni g \mapsto \int_E k(\cdot, y)g(y)d\mu(y) \in H$.
4. A random element in an RKHS $H(k)$ with a measurable reproducing kernel k is a stochastic process, and a stochastic process that takes values in RKHS $H(k)$ is a random element of $H(k)$.

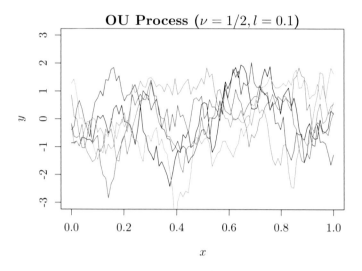

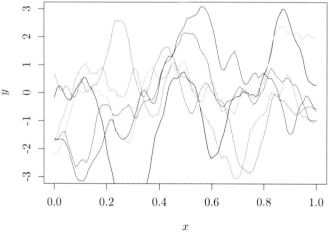

Fig. 6.5 The Orstein-Uhlenbeck process ($\nu = 1.2$, top) and the Matérn process ($\nu = 3/2$, top) for $l = 0.1$

For the proof, see Chap. 7 in [14].

In this section, we learn the properties of random elements and apply them to functional data analysis [24].

First, we consider the average $\mathbb{E}[\langle F, g \rangle]$ of $\langle F, g \rangle$ for each $g \in H$ under $\mathbb{E}[\|F\|] < \infty$. Since $g \mapsto \mathbb{E}[\langle F, g \rangle]$ is a linear functional, there exists a unique $m \in H$ such that

$$\mathbb{E}[\langle F, g \rangle] = \langle m, g \rangle \tag{6.36}$$

from Proposition 22. We write this formally as $m = \mathbb{E}[F]$, which is the definition of the mean of a random element F.

Proposition 71 If $\mathbb{E}\|F\|^2 < \infty$, then

$$\mathbb{E}\|F - m\|^2 = \mathbb{E}\|F\|^2 - \|m\|^2$$

holds.

Proof: If we substitute $g = m$ into (6.36), we obtain

$$\mathbb{E}\|F - m\|^2 = \mathbb{E}\|F\|^2 - 2\mathbb{E}\langle F, m\rangle + \|m\|^2 = \mathbb{E}\|F\|^2 - 2\langle m, m\rangle + \|m\|^2.$$

□

Since $\mathbb{E}\|F\|^2 < \infty$ implies that $\mathbb{E}\|F\| < \infty$, we proceed with our discussion by assuming the former case.

Regarding covariance, if $H = \mathbb{R}^p$, then the covariance matrix is

$$\mathbb{E}[(F - \mathbb{E}[F])(F - \mathbb{E}[F])^\top] = \mathbb{E}[(F - \mathbb{E}[F]) \otimes (F - \mathbb{E}[F])] \in \mathbb{R}^{p \times p}$$

for $F \in \mathbb{R}^p$. For the general Hilbert space H, the correspondence

$$H^2 \ni (g, h) \mapsto \mathbb{E}[\langle F - m, g\rangle\langle F - m, h\rangle] \in \mathbb{R}$$

is linear for each of g and h. Moreover, if $\mathbb{E}\|F\|^2 < \infty$, then it is bounded from

$$\mathbb{E}[\langle F - m, g\rangle\langle F - m, h\rangle] \le \mathbb{E}\|F - m\|^2 \cdot \|g\| \, \|h\| \le \mathbb{E}\|F\|^2 \cdot \|g\| \, \|h\|.$$

If we define $u \otimes v \in B(H)$ by

$$H \ni w \mapsto (u \otimes v)w = \langle u, w\rangle v \in H$$

for $u, v, w \in H$, then a $K \in B(H)$ exists such that

$$\mathbb{E}[\langle F - m, g\rangle\langle F - m, h\rangle] = \mathbb{E}[\langle\{(F - m) \otimes (F - m)\}g, h\rangle] = \langle Kg, h\rangle = \langle g, K^*h\rangle.$$

If we exchange g, h, we obtain the same value, so K and K^* coincide, and each is self-conjugated. Such a K is called a covariance operator, and we formally write this as $K = \mathbb{E}[(F - m) \otimes (F - m)]$.

Proposition 72

$$\mathbb{E}[(F - m) \otimes (F - m)] = \mathbb{E}[F \otimes F] - m \otimes m$$

Proof: From (6.36), for arbitrary $g, h \in H$, we have

$$\mathbb{E}[\langle m, g\rangle\langle F, h\rangle] = \langle m, g\rangle\langle\mathbb{E}[F], h\rangle = \langle m, g\rangle\langle m, h\rangle = \langle(m \otimes m)g, h\rangle.$$

From

$$\langle F - m, g \rangle \langle F - m, h \rangle = \langle F, g \rangle \langle F, h \rangle - \langle F, g \rangle \langle m, h \rangle - \langle F, g \rangle \langle m, h \rangle + \langle m, g \rangle \langle m, h \rangle$$

we have

$$\mathbb{E}[\langle \{(F - m) \otimes (F - m)\}g, h \rangle] = \mathbb{E}[\langle \{F \otimes F - m \otimes m\}g, h \rangle].$$

$\square$

In the following, for simplicity, we proceed with our discussion by assuming that $m = 0$.

Proposition 73 *If $m = 0$ and $\mathbb{E}\|F\|^2 < \infty$, then*

1. *The covariance operator K is nonnegative definite and is a trace class operator whose trace is*
$$\|K\|_{TR} = \mathbb{E}\|F\|^2.$$

2. *With probability 1, $F \in \overline{\mathrm{Im}(K)}$ holds.*

Proof: For $g \in H$,

$$\langle Kg, g \rangle = \mathbb{E}[\langle F, g \rangle \langle F, g \rangle] \geq 0$$

holds, which means that K is nonnegative definite. Moreover, if $\{e_j\}$ is an orthonormal basis, we have

$$\|K\|_{TR} = \sum_{j=1}^{\infty} \langle Ke_j, e_j \rangle = \sum_{j=1}^{\infty} \langle \mathbb{E}[F \otimes F]e_j, e_j \rangle = \mathbb{E}\|F\|^2 < \infty.$$

For the second item, note that in general, we have

$$(\mathrm{Im}(K))^{\perp} = \mathrm{Ker}(K). \tag{6.37}$$

In fact, if we set $g \in \mathrm{Ker}(K)$, then K is self-adjoint $(K = K^*)$ and

$$\langle g, Kh \rangle = \langle Kg, h \rangle = 0 , \ h \in H.$$

Therefore, g is orthogonal to any element of $Im(K)$, and we require $g \in \mathrm{Im}(K)^{\perp}$. Conversely, if $g \in \mathrm{Im}(K)^{\perp}$, then $KKg \in \mathrm{Im}(K)$ and $\|Kg\| = \langle g, KKg \rangle = 0$, i.e., we have $g \in \mathrm{Ker}(K)$. This means that for $g \in \mathrm{Im}(K)^{\perp}$, we have $\mathbb{E}[\langle F, g \rangle^2] = \langle Kg, g \rangle = 0$, and F is orthogonal to any $g \in \mathrm{Im}(K)^{\perp}$ with a probability of 1. Therefore, from Proposition 20, with a probability of 1, we have

$$F \in (\mathrm{Im}(K))^{\perp\perp} = \overline{\mathrm{Im}(K)}.$$

$\square$

Additionally, from Propositions 27 and 31 and the first item of Proposition 73, the following holds.

Proposition 74 *The eigenvalue function $\{e_j\}$ of the covariance operator K is an orthonormal basis of $\overline{\mathrm{Im}(K)}$; the corresponding eigenvalues $\{\lambda_j\}_{j=1}^{\infty}$ are nonnegative, monotonically decrease, and converge to 0. Furthermore, the multitude of each of the nonzero eigenvalues is finite.*

Additionally, from Propositions 73 and 74, the following holds.

Proposition 75 *If $\{f_j\}$ is an orthonormal basis of H, then we have*

$$\mathbb{E}\|F - \sum_{j=1}^{n}\langle F, f_j\rangle f_j\|^2 = \mathbb{E}\|F\|^2 - \sum_{j=1}^{n}\langle Kf_j, f_j\rangle \tag{6.38}$$

which is minimized when $f_j = e_j$ $(1 \le j \le n)$.

Proof: The following two equations imply (6.38).

$$\mathbb{E}\|F - \sum_{j=1}^{n}\langle F, f_j\rangle f_j\|^2 = \mathbb{E}\|F\|^2 + \mathbb{E}\|\sum_{j=1}^{n}\langle F, f_j\rangle f_j\|^2 - 2\mathbb{E}\left[\langle F, \sum_{j=1}^{n}\langle F, f_j\rangle f_j\rangle\right]$$

$$\mathbb{E}\|\sum_{j=1}^{n}\langle F, f_j\rangle f_j\|^2 = \mathbb{E}\left[\langle F, \sum_{j=1}^{n}\langle F, f_j\rangle f_j\rangle\right] = \sum_{j=1}^{n}\mathbb{E}\left[\langle F, f_j\rangle^2\right] = \sum_{j=1}^{n}\langle Kf_j, f_j\rangle.$$

Then, from $\mathbb{E}\|F\|^2 = \|K\|_{TR} = \sum_{j=1}^{\infty}\lambda_j$ (Proposition 73) and Proposition 28, we obtain Proposition 75. $\qquad\square$

For example, from the independent realizations $F_1, \ldots, F_N$ of the random element F, via

$$m_N = \frac{1}{N}\sum_{i=1}^{N}F_i \tag{6.39}$$

$$K_N = \frac{1}{N}\sum_{i=1}^{N}(F_i - m_N) \otimes (F_i - m_N) \tag{6.40}$$

we can estimate the mean m and covariance operator[1] K.

In the following, we examine how to perform principal component analysis (PCA) based on functional data analysis [24].

Then, to obtain the eigenfunctions and eigenvalues, for $x_1, \ldots, x_n \in E$, $1 \le n \le N$, $F_i : E \to \mathbb{R}$, we apply the ordinary (nonfunctional) PCA approach to $X = (F_i(x_k))$ $(i = 1, \ldots, N$ and $k = 1, \ldots, n)$.

---

[1] The denominator of K_N may be $N - 1$.

1. Prepare the basis function $\eta = [\eta_1, \ldots, \eta_m]^\top : E \to \mathbb{R}^m$.
2. Calculate $W = (w_{i,j}) = \int_E \eta(x)\eta(x)^\top dx$ such that $W = (w_{i,j}) = \int_E \eta_i(x)$ $\eta_j(x)dx$.
3. Find $C = (c_{i,j})_{i=1,\ldots,N, j=1,\ldots,m} \in \mathbb{R}^{N \times m}$ such that $F_i(x) = \sum_{j=1}^m c_i^\top \eta_j(x)$.
4. Find the coefficients $d_1, \ldots, d_m$ of the estimated mean function $m_N(x) :=$ $\frac{1}{N}\sum_{i=1}^N F_i(x)$ $(m_N(x) = \sum_{j=1}^m d_j^\top \eta_j(x))$.
5. Since the variance function is

$$k(x, y) = \frac{1}{N}\sum_{i=1}^N \{F_i(x) - m_N(x)\}\{F_i(y) - m_N(y)\} = \frac{1}{N}\eta(x)^T (C - d)^\top (C - d)\eta(y)$$

if we set the eigenvectors as $\phi(x) = b^\top \eta(x)$ $(b \in \mathbb{R}^m)$, then the eigenvalue problem for the covariance operator

$$\int_E k(x, y)\phi(y)dy = \lambda\phi(x)$$

under $b^\top Wb = 1$ reduces to the problem of finding a b such that

$$\eta(x)^\top \frac{1}{N}(C - d)^\top (C - d)\eta(x)\eta(x)^\top b = \lambda\eta(x)^\top b$$

which is equivalent to

$$\frac{1}{N}(C - d)^\top (C - d)Wb = \lambda b.$$

In particular, if we set $u := W^{1/2}b$, it becomes the problem of finding a $u \in \mathbb{R}^m$ such that

$$\frac{1}{N}W^{1/2}(C - d)^\top (C - d)W^{1/2}u = \lambda u$$

under $\|u\| = 1$.

Example 92 For $E = [-\pi, \pi]$, if we set

$$\eta_j(x) = \begin{cases} \frac{1}{\sqrt{2\pi}}, & j = 1 \\ \frac{1}{\sqrt{\pi}}\cos kx, & j = 2k \\ \frac{1}{\sqrt{\pi}}\sin kx, & j = 2k + 1 \end{cases}$$

we have

$$\int_{-\pi}^{\pi} \eta_i(x)\eta_j(x)dx = \delta_{i,j},$$

and W is the unit matrix of size p. Therefore, the eigenequation becomes $\frac{1}{n}(C - d)^\top (C - d)u = \lambda u$, and we can apply $C \in \mathbb{R}^{n \times p}$ instead of the design matrix to the PCA procedure (even if we set $d = 0$ in the above procedure, the centering step will be completed automatically). In this example, we apply Canadian weather data from the `fda` package containing a daily list of, the temperature and precipitation for each day of the year in each Canadian city. We construct the following programs in various ways. We do not give the n functions from the beginning but from $N = 365$ days, as this represents the change in temperature by a linear sum of p bases (Fourie transformation). Therefore, we can say that the function is discretized using a sufficiently large p.

```
library(fda)
g=function(j,x){            ## Basis consisting of p elements
  if(j==1) return(1/sqrt(2*pi))
  if(j%%2==0) return(cos((j%/%2)*x)/sqrt(pi))
  else return(sin((j%/%2)*x)/sqrt(pi))
}
beta=function(x,y){         ## Coefficients in front of the p elements
    X=matrix(0,N,p)
    for(i in 1:N)for(j in 1:p)X[i,j]=g(j,x[i])
    beta=solve(t(X)%*%X+0.0001*diag(p))%*%t(X)%*%y
    return(drop(beta))
}
N=365; n=35; m=5; p=100; df=daily
C=matrix(0,n,p)
for(i in 1:n){x=(1:N)*(2*pi/N)-pi; y=as.vector(df[[2]][,i]); C[i,]=
    beta(x,y)}
res=prcomp(C)
B=res$rotation
xx=res$x
```

Each line of $C \in \mathbb{R}^{n \times p}$ is the coefficient (p) of a function. Then, $B \in \mathbb{R}^{p \times m}$ ($m \le p$) is the principal component vector, and xx is the score of each function. The m-th column vector of B is the vector of m principal components (the coefficients in front of $\eta_j(x)$ fpr $j = 1, \ldots, p$). We change m and the function z and run the following program to see if we can recover the original function.

```
z=function(i,m,x){      ## The approximated function using m components rather than
    m
    S=0
    for(j in 1:p)for(k in 1:m)for(r in 1:p)S=S+C[i,j]*B[j,k]*B[r,k]*g(
        r,x)
    return(S)
}
x.seq=seq(-pi,pi,2*pi/100)
plot(0,xlim=c(-pi,pi),ylim=c(-15,25),type="n",xlab="Days",ylab="
    Temp(C)",
main="Reconstruction for each m")
lines(x,df[[2]][,14],lwd=2)
for(m in 2:6){
    lines(x.seq,z(14,m,x.seq),col=m,lty=1)
}
legend("bottom",legend=c("Original",paste("m=",2:6)), lwd=c(2,rep
    (1,5)), col=1:6,ncol=2)
```

Reconstructions for $m = 2, 3, 4, 5, 6$

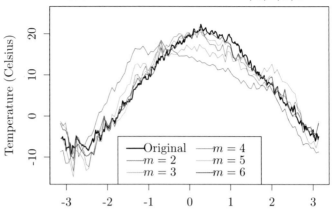

Dates (trandformed from Jan. 1 through Dec. 31 to $-\pi$ through π)

Fig. 6.6 We present the output of approximating Toronto's annual temperature by using $m = 2, 3, 4, 5, 6$ principal components. As m increases, the data are faithfully recovered from the original data

Fig. 6.7 Contribution of temperature to Canadian weather. We can calculate the contribution rate as in the ordinary case where functional data analysis is not used

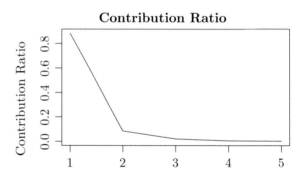

Figure 6.6 shows the output of approximating the annual temperature in Toronto by using $m = 2, 3, 4, 5, 6$ principal components.

Next, we list the principal components in order of increasing eigenvalue and draw a graph of their contribution ratio (Fig. 6.7).

```
lambda=res$sdev^2
ratio=lambda/sum(lambda)
plot(1:5,ratio[1:5],xlab="PC1 through PC5",ylab="Ratio",type="l",
    main="Ratio ")
```

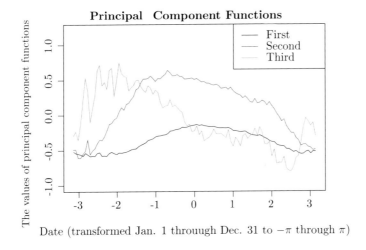

Fig. 6.8 The first, second, and third principal component functions for temperature in the Canadian weather data. Some of the principal component functions are multiplied by -1, which means that they are upside down compared to those of other packages. Additionally, because the horizontal axis is normalized by $[-\pi, \pi]$, the value of each eigenfunction is multiplied by $\sqrt{365/(2\pi)}$

The principal component function is a function with the principal component vector as the coefficients of the basis. It differs from the output of the fda package in two ways.

1. Because the dates of the year are normalized from January 1-December 31 to $[-\pi, \pi]$, the value of the principal component function is multiplied by $\sqrt{365/(2\pi)}$, and the score function is multiplied by $\sqrt{2\pi/365}$.
2. Some principal component vectors are multiplied by -1, resulting in an upside-down function approximation (which is unavoidable if the packages are different).

The first, second, and third principal component functions appear as shown in Fig. 6.8. We use the following program. The first principal component is the effect for the whole year, with winter temperatures influencing the variations between cities.

```
h=function(coef,x){        ## Define a function using coefficients
  S=0
  for(j in 1:p)S=S+coef[j]*g(j,x)
  return(S)
}
plot(0,xlim=c(-pi,pi),ylim=c(-1,1),type="n")
for(j in 1:3)lines(x.seq,h(B[,j],x.seq),col=j)
```

Finally, we display the scores of the 35 Canadian cities (Fig. 6.9).

```
index=c(10,12,13,14,17,24,26,27)
others=setdiff(1:35,index)
first=substring(df[[1]][index],1,1)
```

```
4  plot(0,xlim=c(-25,35),ylim=c(-15,10),type="n",xlab="PC1",ylab="PC2"
       ,main="Canadian Weather")
5  points(xx[others,1],xx[others,2],pch=4)
6  points(xx[index,1], xx[index,2], pch = first, cex = 1, col =rainbow
       (8))
7  legend("bottom",legend=df[[1]][index], pch=first, col=rainbow(8),
       ncol=2)
```

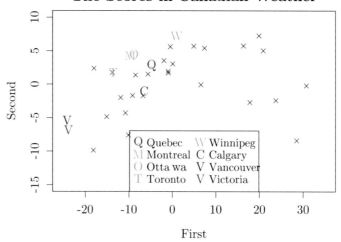

Fig. 6.9 Canadian weather temperature scores, with warmer regions such as Vancouver and Victoria appearing furthest to the left in the first principal component

Appendix

Proof of Proposition 66

Proof: Since the expectation and variance of $f(\omega, x) - m(x)$ are 0 and $k(x, x)$, respectively, and the covariance between $f(\omega, x) - m(x)$ and $f(\omega, y) - m(y)$ is $k(x, y)$, from

$$
\begin{aligned}
&\mathbb{E}[|f(\omega, x) - f(\omega, y)|^2] \\
&= \mathbb{E}[(\{f(\omega, x) - m(x)\} - \{f(\omega, y) - m(y)\} - \{m(x) - m(y)\})^2] \\
&= k(x, x) + k(y, y) - 2k(x, y) + \{m(x) - m(y)\}^2,
\end{aligned}
\tag{6.41}
$$

The continuity of m, k implies (6.30). Conversely, if we assume (6.30), then the continuity of m is obtained from

$$|m(x) - m(y)| = |\mathbb{E}[f(\omega, x) - f(\omega, y)]| \le \{\mathbb{E}[|f(\omega, x) - f(\omega, y)|^2]\}^{1/2}.$$

Without loss of generality, if we assume that $m \equiv 0$, then we have

$$k(x, y) - k(x', y') = \{k(x, y) - k(x', y)\} + \{k(x', y) - k(x', y')\},$$

and each of the right-hand side terms are bounded by

$$|k(x, y) - k(x', y)| = |\mathbb{E}[f(\omega, x)f(\omega, y)] - \mathbb{E}[f(\omega, x')f(\omega, y)]|$$
$$\le \mathbb{E}[f(\omega, y)^2]^{1/2}\mathbb{E}[\{f(\omega, x) - f(\omega, x')\}^2]^{1/2} = \{k(y, y)\}^{1/2}\{\mathbb{E}[|f(\omega, x) - f(\omega, x')|^2]\}^{1/2}$$

and

$$|k(x', y) - k(x', y')| \le \{k(x', x')\}^{1/2}\{\mathbb{E}[|f(\omega, y) - f(\omega, y')|^2]\}^{1/2}.$$

Thus, we have established the continuity of k. □

Proof of Proposition 67

We define $I_f^{(n)}(g) := I_f(g; \{(E_i, x_i)\}_{1 \le i \le M(n)})$. Then, we have $\mathbb{E}[I_f^{(n)}(g)] = 0$. From $\mathbb{E}[f(\omega, x)] = 0$, $x \in E$, and the convergence proven thus far, we obtain the first claim:

$$|\mathbb{E}[I_f(\omega, g)]| = |\mathbb{E}[I_f(\omega, g) - I_f^{(n)}(g)]| \le \{\mathbb{E}[\{I_f(\omega, g) - I_f^{(n)}(g)\}^2]\}^{1/2} \to 0$$

as $n \to \infty$. From the uniform continuity of k, we obtain the second claim:

$$\left|\mathbb{E}[I_f(\omega, g)f(\omega, x)] - \int_E k(x, y)g(y)d\mu(y)\right|$$
$$\le \left|\mathbb{E}[\{I_f(\omega, g) - I_f^{(n)}(g)\}f(\omega, x)]\right| + |\mathbb{E}[I_f^{(n)}(g)f(\omega, x) - \int_E k(x, y)g(y)d\mu(y)]|$$
$$\le |\{\mathbb{E}[\{I_f(\omega, g) - I_f^{(n)}(g)\}^2]\}^{1/2}\{\mathbb{E}[f(\omega, x)^2]\}^{1/2}$$
$$+ |\sum_{i=1}^{M(n)} \int_{E_i} |k(x, x_i) - k(x, y)|g(y)d\mu(y)| \to 0$$

as $n \to \infty$. From

$$\mathbb{E}[I_f^{(n)}(g)I_f^{(n)}(h)] = \sum_{i=1}^{M(n)}\sum_{j=1}^{M(n)} k(x_i, x_j) \int_{E_i} g(x)d\mu(x) \int_{E_j} h(y)d\mu(y)$$
$$\to \int_E \int_E k(x, y)g(x)h(y)d\mu(x)d\mu(y)$$

we obtain the third claim:

$$\left| \mathbb{E}[I_f(\omega, g) I_f(\omega, h)] - \int_E \int_E k(x, y) g(x) h(y) d\mu(x) d\mu(y) \right|$$

$$\leq \left| \mathbb{E}[\{I_f(\omega, g) - I_f^{(n)}(g)\}\{I_f(\omega, h) - I_f^{(n)}(h)\} + \{I_f(\omega, g) - I_f^{(n)}(g)\} I_f^{(n)}(h) \right.$$

$$\left. + \{I_f(\omega, h) - I_f^{(n)}(h)\} I_f^{(n)}(g)] \right.$$

$$+ \left| \mathbb{E}[I_f^{(n)}(g) I_f^{(n)}(h)] - \int_E \int_E k(x, y) g(x) h(y) d\mu(x) d\mu(y) \right|$$

$$\leq \left| (\mathbb{E}[\{I_f(\omega, g) - I_f^{(n)}(g)\}^2])^{1/2} (\mathbb{E}[\{I_f(\omega, h) - I_f^{(n)}(h)\}^2])^{1/2} \right.$$

$$+ (\mathbb{E}[\{I_f(\omega, g) - I_f^{(n)}(g)\}^2])^{1/2} (\mathbb{E}[I_f^{(n)}(h)^2])^{1/2}$$

$$+ (\mathbb{E}[\{I_f(\omega, h) - I_f^{(n)}(h)\}^2])^{1/2} (\mathbb{E}[I_f^{(n)}(g)^2])^{1/2}$$

$$+ \sum_i \sum_j \int_{E_i} \int_{E_j} |k(x, y) - k(x_i, x_j)| g(x) h(y) d\mu(x) d\mu(y) \to 0.$$

$\square$

Exercises 83~100

84. Construct a function `gp.sample` that generates random numbers $f(\omega, x_1)$, $\ldots, f(\omega, x_N)$ from the mean function m, the covariance function k, and $x_1, \ldots, x_N \in E$ for a set E. Then, set m, k to generate 100 random numbers and examine if the covariance matrix matches the m, k.

85. Using Proposition 61, prove (6.3) and (6.4).

86. In the following program, other than the Cholesky decomposition, is there any step that requires a calculation with $O(N^3)$ complexity?

```
gp.2=function(x.pred){
  h=array(dim=n); for(i in 1:n)h[i]=k(x.pred,x[i])
  L=chol(K+sigma.2*diag(n))
  alpha=solve(L,solve(t(L),y-mu(x)))
  mm=mu(x.pred)+sum(t(h)*alpha)
  gamma=solve(t(L),h)
  ss=k(x.pred,x.pred)-sum(gamma^2)
  return(list(mm=mm,ss=ss))
}
```

87. Show from (6.5) that the negated log-likelihood of $x_1, \ldots, x_N \in \mathbb{R}^p$, $y_1, \ldots$, $y_N \in \{-1, 1\}$ is

$$\sum_{i=1}^N \log[1 + \exp\{-y_i f(x_i)\}].$$

88. Explain that Lines 19 through 24 of the program in Example 86 are used to update $f_X \leftarrow (W + k_{XX}^{-1})^{-1}(W f_X + u)$.

89. Replace the first 100 Iris data (50 Setosa, 50 Versicolor) with the 51st–150th data (50 Versicolor, 50 Versinica) in Example 86 to execute the program.

90. In the proof of Proposition 65, why is it acceptable to replace f_Z in (6.16) of the generation process by $\mu_{f_Z|Y}$ to $\mu(x)$? In $\sigma^2(x)$, the variations due to $f_Z|Y$ and $f(x)|f_Z$ are independent. Why can we assume that they are independent?

91. In Example 88, there is a step in which the function gp.ind that realizes the inducing variable method avoids processing $O(N^3)$ calculations. Where is this step?

92. Show that a stochastic process is a mean-square continuous process if and only if its mean and covariance functions are continuous.

93. From Mercer's theorem (6.31) and Proposition 67, derive Karhunen-Lóeve's theorem. Additionally, for $n = 10$, generate five sample paths of Brownian motion.

94. From the formula for the Matérn kernel (6.35), derive $\varphi_{5/2}$ and $\varphi_{3/2}$. Additionally, illustrate the value of the Matérn kernel ($v = 1, \ldots, 10$) for $l = 0.05$, as in Fig. 6.4.

95. Illustrate the sample path of the Matérn kernel with $v = 5/2, l = 0.1$.

96. Give an example of a random element that does not involve a stochastic process and an example of a stochastic process that does not involve a random element.

97. Prepare a basis function $\eta = [\eta_1, \ldots, \eta_p] : E \to \mathbb{R}^p$ and construct a procedure to find $m_N(x)$ in (6.39). Then, input the Canadian weather data for $N = 35$ and output the result. Additionally, construct a procedure to find $K_N(x)$ in (6.40) and output it as a matrix of size $p \times p$.

98. Suppose that we prepare p basis functions as

$$\{\frac{1}{\sqrt{2\pi}}, \frac{\cos x}{\sqrt{\pi}}, \frac{\sin x}{\sqrt{\pi}}, \frac{\cos 2x}{\sqrt{\pi}}, \frac{\sin 2x}{\sqrt{\pi}}, \cdots\}$$

for $E = [-\pi, \pi]$. Why is $W = (w_{i,j}) = \int_E \eta(x)\eta(x)^\top dx$ a unit matrix?

99. Using the Canadian weather data daily as daily[[3]] (precipitation for each day of the year) instead of daily[[2]] (temperature for each day of the year), find the principal component functions and eigenvalues and output graphs similar to those in Figs. 6.8 and 6.9.

100. Using the fda package, find the principal component functions and eigenvalues for both temperature and precipitation for each day of the year and output graphs similar to those in Figs. 6.8 and 6.9.

Bibliography

1. N. Aronszajn, Theory of reproducing kernels. Trans. Am. Math. Soc. **68**, 337–404 (1950)
2. H. Avron, M. Kapralov, C. Musco, A. Velingker, and A. Zandieh. Random fourier features for kernel ridge regression: Approximation bounds and statistical guarantees. *ArXiv*, abs/1804.09893, 2017
3. C. Baker. *The Numerical Treatment of Integral Equations* (Claredon Press, 1978)
4. P. Bartlett and S. Mendelson. Rademacher and gaussian complexities: Risk bounds and structural results. In *J. Mach. Learn. Res.*, 2001
5. K.P. Chwialkowski and A. Gretton. A kernel independence test for random processes. In *ICML*, 2014
6. R. Dudley. *Real Analysis and Probability*. Cambridge Studies in Advanced Mathematics, 1989
7. K. Fukumizu. *Introduction to Kernel Methods (kaneru hou nyuumon)* (Asakura, 2010). (In Japanese)
8. T. Gneiting, Compactly supported correlation functions. J. Multivar. Anal. **83**, 493–508 (2002)
9. G.H. Golub, C.F. Van Loan, *Matrix Computations*, 3rd edn. (Johns Hopkins, Baltimore, 1996)
10. I.S. Gradshteyn, I.M. Ryzhik, R.H. Romer, Tables of integrals, series, and products. Am. J. Phys. **56**, 958–958 (1988)
11. A. Gretton, K. Borgwardt, M. Rasch, B. Schölkopf, A. Smola, A kernel two-sample test. J. Mach. Learn. Res. **13**, 723–773 (2012)
12. A. Gretton, R. Herbrich, A. Smola, O. Bousquet, B. Schölkopf, Kernel methods for measuring independence. J. Mach. Learn. Res. **6**, 2075–2129 (2005)
13. D. Haussler. Convolution kernels on discrete structures. Technical Report UCSC-CRL-99-10, UCSC, 1999
14. T. Hsing and R. Eubank. *Theoretical Foundations of Functional Data Analysis, with an Introduction to Linear Operators* (Wiley, 2015)
15. K. Itō. *An Introduction to Probability Theory* (Cambridge University Press, 1984)
16. Y. Kano and S. Shimizu. Causal inference using nonnormality. In *Proceedings of the Annual Meeting of the Behaviormetric Society of Japan 47*, 2004
17. K. Karhunen, Über lineare methoden in der wahrscheinlichkeitsrechnung. Ann. Acad. Sci. Fennicae. Ser. A. I. Math.-Phys. **37**, 1–79 (1947)
18. K. Karhunen. *Probability theory. Vol. II* (Springer, 1978)
19. H. Kashima, K. Tsuda, and A. Inokuchi. Marginalized kernels between labeled graphs. In *ICML*, 2003
20. S. Lauritzen. *Graphical Models* (Oxford Science Publications, 1996)

© The Editor(s) (if applicable) and The Author(s), under
exclusive license to Springer Nature Singapore Pte Ltd. 2022
J. Suzuki et al., *Kernel Methods for Machine Learning with Math and R*,
https://doi.org/10.1007/978-981-19-0398-4

21. J. Mercer. Functions of positive and negative type and their connection with the theory of integral equations. *Philosophical Transactions of the Royal Society A*, pp. 441–458, 1909
22. J. Neveu, *Processus aleatoires gaussiens, Seminaire Math. Sup.* Les presses de l'Universite de Montreal, 1968
23. A. Rahimi and B. Recht, Random features for large-scale kernel machines. In *Advances in neural information processing systems*, 2007
24. J. Ramsay and B.W. Silverman. *Functional Data Analysis* (Springer Series in Statistics, 2005)
25. C. Rasmussen and C.K.I. Williams. *Gaussian Processes for Machine Learning* (MIT Press, 2006)
26. B. Schölkopf, A. Smola and K. Müller. Kernel principal component analysis, In *ICANN*, 1997
27. R. Serfling. *Approximation Theorems of Mathematical Statistics* (Wiley, 1980)
28. S. Shimizu, P. Hoyer, A. Hyvärinen, A.J. Kerminen, A linear non-gaussian acyclic model for causal discovery. J. Mach. Learn. Res. **7**, 2003–2030 (2006)
29. I. Steinwart, On the influence of the kernel on the consistency of support vector machines. J. Mach. Learn. Res. **2**, 67–93 (2001)
30. M.H. Stone, Applications of the theory of boolean rings to general topology. Trans. Am. Math. Soc. **41**(3), 375–481 (1937)
31. M.H. Stone, The generalized weierstrass approximation theorem. Math. Mag. **21**(4), 167–184 (1948)
32. K. Tsuda, T. Kin, K. Asai, Marginalized kernels for biological sequences. Bioinformatics **18**(Suppl 1), S268-75 (2002)
33. J.-P. Vert. *Aronszajn's theorem*, 2017. https://members.cbio.mines-paristech.fr/~jvert/svn/kernelcourse/notes/aronszajn.pdf
34. K. Weierstrass. Über die analytische darstellbarkeit sogenannter willkürlicher functionen einer reellen veränderlichen. *Sitzungsberichte der Königlich Preußischen Akademie der Wissenschaften zu Berlin*, pp. 633–639, 1885. Erste Mitteilung
35. K. Weierstrass. Über die analytische darstellbarkeit sogenannter willkürlicher functionen einer reellen veränderlichen. *Sitzungsberichte der Königlich Preußischen Akademie der Wissenschaften zu Berlin*, pp. 789–805, 1885. Zweite Mitteilung
36. H. Zhu, C.K.I. Williams, R. Rohwer and M. Morciniec. Gaussian regression and optimal finite dimensional linear models. In *Neural Networks and Machine Learning*, 1997

Printed in the United States
by Baker & Taylor Publisher Services